AUTOPIA

THE FUTURE OF CARS

'An interesting read... thorough enough to delight even those who consider themselves to be on the ball with industry developments.'
Auto Express

'A riveting and quirky look at the latest in car technology. Perfect for geeks, petrolheads and geeky petrolheads.'
Rick Edwards, presenter of *!mpossible*

'Jon is a passionate, articulate and characteristic car man who has a canny knack of capturing the very essence of the sector I love so much! Exactly what he has done with *Autopia*. Brilliant.'
Ant Anstead, presenter of Channel 4's *For the Love of Cars*

'Entertaining, thought-provoking and well researched, *Autopia* is an essential read for anyone interested in the future of cars.'
Jason Dawe, former presenter of *Top Gear*

'Intelligent, honest and extremely interesting. Just like Jon.'
Richard Porter, script editor of *The Grand Tour*

Jon Bentley is a presenter on Channel 5's *The Gadget Show*. He was the producer and executive producer of *Top Gear* for many years and has a bend named after him on the programme's test track. He is a committed car enthusiast and has written for multiple car publications.

AUTOPIA

THE FUTURE OF CARS

JON BENTLEY

Atlantic Books
London

First published in hardback in Great Britain in 2019 by Atlantic Books,
an imprint of Atlantic Books Ltd.

This edition published in 2020

10 9 8 7 6 5 4 3 2 1

A CIP catalogue record for this book is available from the British Library.

Paperback ISBN: 978 1 78649 635 5
E-book ISBN: 978 1 78649 636 2

Printed in Great Britain

Atlantic Books
An imprint of Atlantic Books Ltd
Ormond House
26–27 Boswell Street
London
WC1N 3JZ

www.atlantic-books.co.uk

CONTENTS

INTRODUCTION

WHY THE CAR MATTERS

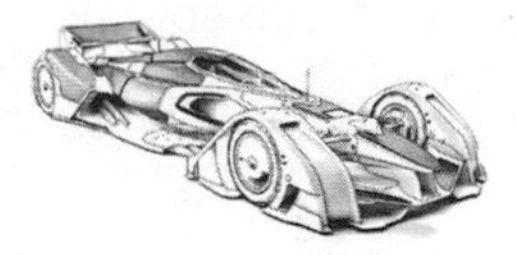

Since its invention nearly 130 years ago the car has been an extraordinary global success story, extending its influence over most of the planet and transforming vast tranches of human life. There are now about 1.3 billion of them scurrying round the world's roads. By 2035 there'll likely be more than 2 billion – a growth rate comfortably exceeding that of the human population. The car has revolutionised patterns of human settlement, reshaped the geography of cities and countryside and transformed the working lives of hundreds of millions of people.

Cars are so much more than a means of getting from A to B. From the beginning to the end of our lives, they're a powerful psychological force. A few weeks after my seventeenth birthday, I passed my driving test and experienced that amazing sense of freedom that only the car can give. Suddenly I had the means to go wherever I wanted with minimal effort. Like countless drivers before and after me, I embraced an explosion of new opportunities for employment, travel and (yes) romance.

People may say that taking your driving test is no longer the universal rite of passage that it used to be back then in the late 1970s, but I can vividly recall my seventeen-year-old daughter on her return from her first fully independent drive saying it was the most liberating thing she'd ever done. More soberingly, I can remember the look of crestfallen disappointment on my late mother's face when I told her she should give up driving on account of her all-consuming Alzheimer's. That was the end of the road in more ways than one.

As objects go, cars have an incredible capacity to inspire loyalty and affection in their owners. They can rival our pet cat or dog because, like our favourite animal companions, they also have their own loveable quirks. There's the noise of the engine, the sensation of acceleration, the rich variety of dynamic experiences through steering, cornering and handling characteristics plus the tactile sensations of the controls, the ways the doors open and shut, the weight and feel of throttle and clutch. All of these create the potential for a truly satisfying relationship between driver and machine.

Your choice of motor says a lot about you. For a nineties sales rep, embroiled in British company-car culture, fretting over whether the badge on the boot-lid was an L or a GL was practically part of the job description. Boy racers of the era agonised in meticulous detail over the specs of a VW Golf GTI versus a Peugeot 205 GTi. But the psychology isn't solely an individual affair. There's an intense social dimension to car ownership. Banger racers, Vintage Sports-Car Club members, muscle-car enthusiasts and Austin Allegro owners are all part of a developing network of driving tribes. People make friends all over the world through their ownership of a particular type of car.

The last fifty years have witnessed the golden age of the car, with the once-spluttering invention dominating the globe and shaping all our lives. But is this set to change? Along with its power to transform the world for the better, mass car ownership has also had a profoundly negative legacy. Doctors talk of how emergency departments took on a whole new level of gruesome in the 1960s. Previously quiet Saturday nights became bloodfests. Pollution and resource depletion became a concern shortly afterwards and, in recent years, the internal combustion engine has been pilloried for its contributions to climate change. A whole catalogue of lesser problems include ever more congestion, road rage, the loss of rural idylls to commuting and tourism and, let's not forget, the increasing nightmare of finding anywhere to park.

Rather than representing freedom and opportunity, are cars going to become an expensive anachronism that take up too much of our money and time? Are they a burden in an age when urban populations are far more concentrated?

There is reason to hope that all is not lost. Most cars today still have four wheels and run on fossil fuels, but that doesn't mean that we haven't already seen incredible breakthroughs in speed, safety, efficiency and – of course – gadgetry. And, while the technology has remained fundamentally the same for decades, industry insiders insist that we are on the cusp of a revolution.

I'm writing this book to explore such innovations and see where they might lead in the years to come. From self-driving trucks to cars that can see round corners, I'll examine the mind-blowing challenge of successfully designing artificial intelligence to take the place of human drivers. I'll explore the options available for powering the cars of the future and question whether electric cars will ever get over their bugbears of range anxiety and sluggish charging. I'll ask whether hydrogen will at last realise its potential and whether diesel really deserves to be demonised.

By talking to designers young and old, I'll discover the ways in which future cars will retain their allure – and their speed. I'll ask whether cars will continue to get safer or if increased technology will make the car into an unpredictable lethal weapon as hackers bypass their often shamefully lax security. If the car embarks on a radical new career of change, what will happen to all the existing ones – not least those classic cars lovingly cherished by millions of enthusiasts? Finally, I'll consider some of the more unlikely future scenarios for personal transport and if the car will remain relevant in the face of fierce potential competition.

I've been fortunate to spend my whole working life engaged with my two childhood obsessions – cars and technology. I worked with the iconic TV series *Top Gear* for fifteen years, many of those as its producer. Red-letter days came thick and fast. I can remember getting 190 mph out of a Porsche 959 on the autobahn entirely legally near Stuttgart, driving my first 911 – my first Ferrari – and my first drives in a whole spectrum of iconic classics, like an XK120, D-Type and Aston Martin DB4 and 6. And there was the time I drove Alan Clark up and down the Italian Alps in his Rolls-Royce Silver Ghost.

As well as my car passion I can barely recall a time when I wasn't fascinated by gadgetry and technology. I used to build radios, mend TVs and practise old-fashioned analogue photography, including developing and printing, and I have dabbled with computers for as long as they've been around. In recent years I have presented *The Gadget Show*, the UK's most popular consumer-technology programme.

Dr Ian Robertson, BMW's global sales chief, said in June 2017 that 'The car industry has done the same thing for over a hundred years. In the next five to seven years the car will change enormously. We're at the tipping point in an industry worth $2 trillion a year.' It's an incredibly exciting time, but also a crucial one. Avid customers no longer queue round the block for a fresh model, like they did in the 1960s with a Ford Mustang. Instead, they are far more likely to camp outside an Apple store, waiting for the latest iPhone. If the car is to stay relevant, it needs to evolve.

'Connected', 'autonomous' and 'zero emission' are the industry buzzwords right now. According to a designer I spoke to from Jaguar Land Rover, every single one of the future cars in their studios is battery-powered and self-driving. But are our favourite brands on the right track? Or will our love of being behind the wheel win out? Will we really have the patience with battery-powered cars? Certainly it's hard to see us giving up on cars as a whole but, in a world of rapid technological change, it feels like it's time for the car industry to catch up. Let's start with perhaps the most talked about and the most ambitious of the predicted changes: autonomy – or the car that drives itself.

ONE

CONNECTED AND AUTONOMOUS

the Rise of the Robot Cars

The bright-red Golf GTI weaved in and out of cones at the very limits of its grip. We took the sharp corner ahead and I was thrown to the left as the driver took a perfect racing line and began to accelerate out of the turn. I was being hurled round the test track at VW's rather remote facility in Wolfsburg and feeling in awe of the test driver's remarkable command of the course.

What made it more impressive was that the driver wasn't even human. In fact, there was no visible driver at all. The steering wheel, accelerator and brake were all magically moving entirely of their own accord. This wasn't even a Google car: it was 2006, when autonomous aspirations were yet to hit the mainstream. You can understand why I found the effect so stunning.

Car automation has a surprisingly long history. For decades, scientists have sought to slash the death toll on our roads by replacing the fallible human driver with a more capable technological alternative. Until recently such aspirations were confined to science fiction, their real-world potential thwarted by practicalities of technology and cost. But now, thanks to recent improvements in computer power, artificial intelligence, machine learning and sensor technologies, the impossible is becoming possible.

The driverless journey started with a radio-controlled car that hit the streets of New York in 1925. Inventor Francis Houdina fitted a brand-new Chandler with a radio receiver and 'apparatus' attached to the steering column. This turned in response to signals from a radio transmitter in a car following behind. According to a contemporary report in the *New York Times*, the car drove 'as if a phantom hand were at the wheel'.

The initial unveiling didn't go well. After making wildly uncertain progress down Broadway, the car narrowly missed a fire engine and crashed into a car full of news cameras recording the whole operation. Police instructed Houdina to abort the experiment. Even more bizarrely, the similarly named Harry Houdini became irritated by Houdina's efforts and accused him of 'using his name unlawfully in the conduct of business'. The famous magician broke into the Houdina Radio Control Co. and vandalised the place – a misdemeanour for which he was later summoned to court.

The automation journey stuttered on with 'magic motorways', which were first shown at General Motors' 'Futurama'

The autonomy spectrum

There are six levels of automation as defined by the Society of Automotive Engineers:

Level 0 No automation.

Level 1 The most basic level of automation, whereby just one function of the driving process is taken over. The car might have lane centring or adaptive cruise control but not both.

Level 2 In which multiple functions are controlled – both lane centring and adaptive cruise control, for example.

Level 3 So-called 'conditional automation', where the car can take control of safety-critical functions but still needs a driver to be permanently paying attention in case intervention is necessary. A Level 3 car might take over driving in a low-speed traffic jam, for instance.

Level 4 Whereby cars are autonomous but only in controlled areas – say, a robotaxi operating on a housing estate. Level 4 cars do not need steering wheels or pedals. (Some wags have suggested that horses are a Level 4 autonomous vehicle.)

Level 5 The 'fully autonomous' stage. The car can take over completely and doesn't require special lane markings or any other dedicated infrastructure; it really can self-drive anywhere, and the 'driver' can go to sleep or do anything they wish.

exhibit at the 1939 World's Fair in New York. A brainchild of designer Norman Bel Geddes, the concept featured electromagnetic propulsion and guidance systems built into the road. Embedded circuits in the road were also behind experiments to guide cars by the American electronics company RCA. It started with model cars in 1953 and graduated to real ones in 1958. Sensors in the front bumpers picked up signals from a buried cable that provided information on roadworks and stalled cars ahead; the system would apply the brakes or change lanes as required. The company thought self-driving cars would be widespread on highways by 1975. The British government's Road Research Laboratory (later the Transport Research Laboratory, or TRL) came up with a hands-free Citroën DS prototype a year or two later that worked in a similar way – and it too predicted that by the 1970s all motorways would feature a lane offering hands-free driving. Like many who followed, its claims were wildly optimistic.

The first real stand-alone autonomous vehicle appeared in Japan in 1977, but it was far from being really roadworthy. Instead of buried electronics it relied on a huge computer that occupied most of the dashboard and the passenger footwell. Using information gleaned about its environment from inbuilt cameras, it could follow white lines on the tarmac – though only at a rather pedestrian 20 mph. Nevertheless, this was one of the first vehicles to move beyond level 0 on today's autonomy spectrum, as defined by the American organisation SAE International, formerly known as the Society of Automotive Engineers.

German aerospace engineer Ernst Dickmanns upped the levels of speed and artificial intelligence with the help of a boxy Mercedes van. The VaMoRs prototype was tested successfully in 1986 and drove itself at 60 mph on the autobahn a year later. It led the pan-European research organisation EUREKA to launch the painfully named PROgraMme for European Traffic of Highest Efficiency and Unprecedented Safety, or PROMETHEUS project. With a significant injection of €749 million, researchers at the University of Munich developed camera technology, software and computer processing that culminated in two impressive robot vehicles: VaMP and VITA-2, both based on the Mercedes S-Class. In 1994, these piloted themselves accurately through traffic along a 600-mile stretch of highway near Paris at up to 80 mph. A year later, they clocked up 108 mph on a journey from Munich to Copenhagen that included a 98-mile stretch without human assistance.

Many manufacturers started developing limited autonomous features around this time, but they were strictly aimed at driver assistance and certainly couldn't contend with the vast range of hazards we encounter all the time on the road. This would soon change when a new player entered the game: the US military. At the dawn of the twenty-first century, they sponsored the DARPA Grand Challenges, in which a $1 million prize was promised to the team of engineers whose vehicle could navigate itself fastest around a 150-mile obstacle course. Although no vehicles finished the inaugural event in 2004, it generated hype and helped spur innovation. Five vehicles finished the next year's challenge, with a team from Stanford nabbing the $2 million prize.

The Stanford team caught the eye of a certain technology company called Google and the rest is history. In 2010, Google announced that it had been secretly developing and testing a self-driving car system with the aim of cutting the number of car crashes in half. The project, which would later be renamed Waymo, was headed by Sebastian Thrun, director of the Stanford Artificial Intelligence Laboratory, and its goal was to launch a vehicle commercially by 2020.

Six Toyota Priuses and an Audi TT comprised the initial test fleet. Equipped with sensors, cameras, lasers, a special radar and GPS technology, they were completely interactive with their environment rather than restricted to a prescribed test route. The system could detect hazards and identify objects like people, bicycles and other cars at distances of several hundred metres. A test driver was always in the car to take over if necessary.

Google's involvement prompted an explosion of interest in the subject. Investment by established brands in the technology and automotive industries ballooned, along with a bevy of new start-ups. According to American think tank The Brookings Institution, $80 billion was spent on self-driving car attempts between 2014 and 2017. This may prove to be a giant capitalist mistake that'll make the South Sea Bubble, tulip mania and the subprime mortgage meltdown seem positively rational by comparison.

As usual, the targets of when full-scale autonomy would really be achieved were often overly ambitious. It becomes easier to see why when you appreciate how these wonders of technology are actually supposed to work.

Sensing the road

This brave new world of genuinely intelligent cars requires a diverse array of hardware with which the car tries to gain an accurate perception of its environment.

The most expensive, spectacular and distinctive sensors on a self-driving car are LiDAR, which stands for Light Detection and Ranging, usually housed in a roof pod. These systems bounce low-powered invisible laser beams off objects to create extremely detailed and accurate 3D maps of their surroundings. Their field of view can be up to 360 degrees and, because powerful lasers are used, LiDAR has the advantage of working in any lighting conditions.

Scientists have been using lasers to measure distances since the 1960s, when a team from the Massachusetts Institute of Technology (MIT) accurately logged the distance to the moon by measuring how long the light took to travel there and back. Its pioneering use in cars began with an experiment carried out in 2007 by an audio-equipment company called Velodyne. Five vehicles equipped with the company's revolutionary new sensor successfully navigated a simulated urban environment.

Around 2016, LiDAR could cost around $75,000 per car. As of 2019 this sum has fallen to around $7,500 for a top-of-the-range unit. That needs to fall further and Ford is targeting approximately $500 as a cost for the component in the future. At present, most cars use one LiDAR unit, which creates a 360-degree map by either rotating the whole assembly of lasers or by using rapidly spinning mirrors. Many researchers think a key requirement of lowering the cost will be to create solid-state designs with few or no moving parts, eliminating the need for such spinning mechanisms.

Mirrors could possibly be eliminated by so-called phased arrays, which use a row of laser emitters. If they all emit in sync the laser travels in a straight line, but by adjusting the timing of the signals the beam can shift from left to right. Flash LiDAR is another possibility. This operates more like a camera. A single laser beam is diffused to illuminate an entire scene in an instant.

Laser-powered eyes on the road. LiDAR sensors are getting smaller and cheaper.

A grid of tiny sensors then captures the light bouncing back from various directions. It's good because it captures the entire scene in one moment, but it currently results in more noise and less accurate measurement.

There are other stumbling blocks. Once most cars on the road have LiDAR they could soon start interfering with each other. Systems normally fire their lasers in a straight line and use a super accurate clock. They could be easily upset by lasers on other cars operating in the same range. Similarly, sceptics worry about the ability of the system to cope in awful weather. Lastly, to avoid eye damage, the lasers are fairly weak and currently limited in range to about 150 metres. For a car to accelerate and join a stream of fast-moving traffic, the laser range needs to be at least 300 metres. LiDAR manufacturers are working on increasing the laser frequency to allow stronger output with a beam that is further from the visible light range. As the systems improve, it is likely other shortcomings will be dealt with too. The technology already functions decently in snow and rain, and it is getting better at avoiding interference.

While LiDAR allows the car to 'see' over short distances, a different solution is needed for longer distances. This is where radar comes in. Many new cars already have radar sensors, used for adaptive cruise control, blind-spot protection and automatic emergency-braking systems. Their field of view is about 10 degrees and they're relatively cheap at between £80 and £120 per sensor.

Traditionally radar's main advantage is the ability to perceive distance and velocity. It can measure speed from a long way away and it's a well-proven technology. Radar can even see round things. Its wavelengths are relatively long so there's significant diffraction and forward reflection – you can 'see' objects behind other ones. On YouTube there's a video, taken inside a car driving along, which shows radar in action when the car's automatic emergency-braking system suddenly activates and the brakes are applied. The view ahead is showing nothing out of the ordinary; but half a second later the car in front rear-ends the car ahead of it. The car's radar was able to see that the (optically hidden) car two cars ahead had braked suddenly, and then braked hard itself to avoid a crash.

Radar's big disadvantage, and why it needs to be supplemented by other sensors, is that it can't perceive detail. Everything's just a blob. It's no good at distinguishing between a pedestrian and a cyclist even though it can tell whether they're moving or stationary. A Waymo's LiDAR, on the other hand, can not only tell the difference but can also tell which way the pedestrian or cyclist is facing. Ultrasonic sensors are used to measure the position of objects very close to the vehicle too. We're accustomed to them in those bleeping parking sensors.

They were invented in the 1970s and the first volume production car they appeared on was the 2003 Toyota Prius. Their range might be a mere 10 metres or so, but they are very cheap and provide extra essential information in low-speed manoeuvring and about adjacent traffic.

High-resolution video cameras are an important part of a self-driving car's equipment. They are used to recognise things like traffic lights, road markings and street signs – objects that offer visual clues but no depth information. Cameras can also detect colour, which LiDAR can't, and they're better at discerning differences in texture. When in stereo they can also help calculate an object's distance – although this effect diminishes the further away something is, which limits the technique's usefulness in high-speed driving. They are relatively cheap at around £150 per car but they are greatly affected by prevailing light conditions and visibility. Infrared can help here to some extent.

You might think that GPS would also be critical for getting around. However, it is normally accurate to only a few metres and not consistently so, with the signal being easily interrupted by tall buildings and tunnels; its role in the autonomous car is therefore somewhat limited. It can, however, be useful in combination with other sensors. MIT, for example, has built a self-driving Prius that manages its way round back roads pretty well using just GPS, LiDAR and accelerometers.

The brain of the car

Of course, all these sensors would be useless without something to interpret their data. Processing the vast quantity of incoming information – and processing it sharpish – requires a very powerful computer with far more heft than the average PC. Even current cars like Waymos are thought to generate up to 150 gigabytes of data every 30 seconds. That's enough to fill many laptop hard drives and is equivalent to 19 terabytes per hour. The cars also need to store these colossal quantities of information in case of later enquiries, crashes and disputes. This takes its toll on energy consumption, with the typical prototype needing 5,000 watts of power – or about the same as forty typical desktop PCs. That's enough to have a serious hit on fuel consumption or battery range. Imagine switching off the self-driving to conserve the battery and get you to your destination. That wouldn't feel like progress.

There's a lot of effort going into making processors that are more suited to the demands of the self-driving car. Google has developed new chips especially designed for self-driving tasks, called Tensor Processing Units. They fire up only those bits of the chip necessary for a given task, which allows more operations per second and better power efficiency. They're suited to machine learning of many kinds because they can handle a lot of relatively low-power processing tasks at once.

Nvidia, which is best known for making graphics cards on computers, has come up with a chip called Xavier. It allows 30 trillion operations per second (about 500 times more than a decent laptop) on a chip that consumes just 30 watts of power. It packs an amazingly powerful and efficient punch, and it's the most complex system on a chip ever created, with an 8-core CPU and 512-core GPU, a deep learning accelerator, computer vision accelerators, and 8K video processors. 'We're bringing supercomputing from the data centre into the car,' says the bloke leading their automotive work, Danny Shapiro. Still, there's a long way to go until the systems are sufficiently energy- (and cost-) efficient for the mainstream. And you can have the most powerful computers in the world, but without the right software they are just expensive black boxes.

Artificial intelligence

If you thought the hardware was complicated, the software needed to make self-driving cars an everyday reality is hundreds of times more brain-defyingly baffling. This is possibly the greatest test of artificial intelligence the world has ever seen.

The idea is the car takes the various bits of information from all the sensors, and combines them to determine where it is, exactly what surrounds it and how those surroundings will change – and then plots a course of action through the space. This all has to be done within milliseconds and to unfailingly high levels of accuracy.

The car uses a technology called deep learning. I'm oversimplifying, but essentially the on-board computer turns all the information from the sensors into a vast matrix containing billions of bits of digital information. It then searches for known patterns within the data, which it can use to select the right behaviours. This is sometimes termed a neural network. Once patterns and necessary behaviours are detected and decided upon, this is translated into physical form through the accelerator, brakes and steering, as well as by using other systems like the lights and even the horn.

Extra intelligence

Self-driving isn't the only way in which cars will become brainier.

In a straight rip-off from mobile phones, facial recognition could soon be the way you unlock and start your car. Another phone feature coming to your car is a voice assistant. Siri, Alexa and the like will get much better at playing your music and answering life's pressing questions. Electric-car users will increasingly be able to choose what noise their car makes – a sort of ring-tone selection for cars. BMW has already recruited *Blade Runner 2049* composer Hans Zimmer to provide an attractive tone to warn pedestrians of its near-silent cars' presence.

Your new motor will soon be clever enough to receive your deliveries wherever you are. Services will be able to unlock it and leave parcels or even your cleaned laundry inside.

Self-parking will come well before self-driving. Summoning your car a few metres by an app on your smartphone or letting it take over parallel-parking tasks will soon seem very old school. Instead your car will know in advance where available spaces are and navigate itself around a car park to park itself. Then you'll be able to retrieve it automatically when required.

Sound management will get better, with improvements in noise cancellation and vocal enhancement tailored to different seats in the car so you can talk to fellow passengers more easily. Furthermore, increasing proportions of the interior will be plastered with screens so you can customise it with your favourite dashboard style or interior graphics. Augmented reality displays will help streamline GPS guidance and put useful instructions on your windscreen.

In short, the car operates by detecting patterns in data and responding to them. It's the same sort of technology that's helped to predict things ranging from earthquakes to heart disease and has developed the capacity to analyse cancer scans, identify eye disease and muscle damage – all more quickly and accurately than human doctors. It's also proved very useful, if not creepily disturbing, in facial recognition. Episodes like the recording of biometric information about concert-goers when they snap themselves in selfie booths and people being mistakenly arrested for shoplifting merely because they share similar features with criminals have brought facial recognition into some disrepute, and show how deep learning can fall short.

Identifying patterns on the road is far more demanding in terms of the variety of things that need to be recognised and the short time available to do so. Creating software that then knows how to respond to these perceived environments is even more of a nightmare. The most obvious way the car is taught how to behave is through 'behavioural cloning' – accumulating data from how (decent) human drivers behave. Then the system can practice and learn, improving its own driving as it gets better at making decisions itself, while being watched by a real driver in case things go wrong – which they will.

At its most basic, a system might recognise an open road from vast experience of seeing pictures of a clear road ahead, translating that into an action to accelerate up to a predetermined speed limit. At the more challenging end of the scale, it has to cope with a busy, previously un-tackled junction, packed with other cars, trucks, pedestrians and cyclists. It has to predict how they're all going to behave and respond accordingly.

Sebastian Thrun thinks accurate perception is the most difficult challenge. In the early days of Google's autonomous-vehicle project, he recalls, 'our perception module could not distinguish a plastic bag from a flying child'. As already indicated, it's getting better. At a Google conference in 2018, Waymo showed examples of a pedestrian carrying a door-sized plank of wood, a construction worker poking halfway out of a manhole cover and even people in inflatable dinosaur costumes. In each case the pedestrian's profile was obscured but the Waymo car correctly identified them as pedestrians.

There are issues with any deep-learning system. In effect, every situation it faces will need to have been experienced before in some way; otherwise it won't know how to react. One problem is termed 'overfitting'. The system can draw correlations between totally irrelevant attributes. One could imagine trying to predict the score of a dice depending upon the time of day or even its colour. An artificial-intelligence program will always try to construct a hypothesis as to why the scores are occurring based on whatever factors it has at its disposal.

The problem gets worse if more factors are being considered. Imagine I turn left – the system may intuit that I did so because I'm 200 metres from a cyclist and I happen to have done similar several times before when I'm in this part of town at this time of day. This is all a gross oversimplification, but it helps to illustrate the scale of the AI challenges in relation to driving when there are so many variables.

'Underfitting' is the opposite. The AI system can't always capture the correlations we want it to. For example, it might not know where the edge of the road is or perhaps it can't deduce from the laser and camera data that a pedestrian or cyclist is ahead. The usual way to avoid underfitting is to feed the system more data or to give it more experience of correlation between the AI system and the real world. Self-driving car companies have thousands of people manually tagging images with useful information to help avoid underfitting, supplementing neural networks with known real-world data.

'Generalisation' is another problem. If we know what a mouse looks like and what a gerbil looks like, humans can interpret a hamster as something between the two – another mammalian rodent. Artificial intelligence finds this difficult. It struggles to take what it already 'knows' and come up with something different that's sensible. It either doesn't recognise the new object at all or it creates constantly varying descriptions. This is why chatbots really aren't very good at speaking yet. They don't give you the impression they're understanding anything; they're just repeating things they've heard before where they gauge it to be appropriate.

Watching AI at work. The Nvidia chips generate multicoloured oblongs or other confidence-inspiring highlights as they identify familiar shapes like cars, people and bicycles.

Self-driving car software is what AI experts call a 'black box' system. You know what the inputs are. You know what the outputs are. But how the system derives the outputs from the inputs is a mystery. We don't really understand how the algorithms work or how cars 'think'. Nvidia has attempted to visualise this in a self-driving context by highlighting the parts of the image from a car's sensors that are involved in decision-making. Reassuringly the results show that the chips are focusing on the edges of roads, lane markings and parked cars – exactly the things that human drivers would be attending to.

'What's revolutionary about this is that we never directly told the network to care about these things,' Urs Muller, chief architect of Nvidia's self-driving cars, wrote in a blog post.

It isn't a complete explanation of how a neural network reasons, but it's a first step. As Muller says: 'I can't explain everything I need the car to do, but I can show it, and now it can show me what it learned.' There's a strong argument that visualisation should be displayed to passengers in self-driving cars so they can appreciate how the machine is thinking, building confidence in the system.

On top of the neural network, the software suite of a self-driving car normally includes high-definition mapping that needs to be continuously updated over the air. These maps are down to centimetre accuracy (on lane markings, for example), are 3D and have many different layers, including those for real-time activity like traffic lights and so-called 'semantic' features such as road signs.

Who are the main players?

The widespread adoption of self-driving cars in the real world always seems to be tantalisingly close yet just out of reach – like hoverboards and BabelFish-esque real-time translation devices. In 2015, Elon Musk predicted that his company would be selling fully autonomous Teslas by 2018. It wasn't. NuTonomy's plans for driverless taxis in Singapore by 2019 didn't work out either, though trials have begun. GM's steering-wheel-free, fully autonomous production car was due to hit the market in 2019. At the time of writing it hasn't appeared.

Google formed the parent company Alphabet in 2015 and rebranded the Waymo subsidiary the following year. Waymo is widely considered to be way ahead in the self-driving car race, leading by measures such as miles driven autonomously and how many miles its cars cover between 'engagements' when a safety driver in the car is forced to take over control. Some of the most telling statistics come from the California Department of Motor Vehicles (DMV), reflecting the concentration of tech companies and self-driving vehicle experimentation in the state. Waymo declared that its vehicles travelled an average of 11,018 miles between engagements in California in 2018 and celebrated 10 million miles travelled in total by its cars, over a million of which were in 2018 alone.

Since 2015, respected research outfit Navigant has been analysing the sector according to ten criteria: vision, go-to market strategy, partners, production strategy, technology, sales, marketing and distribution, product capability, product quality and reliability, product portfolio and staying power. They put Waymo on top too.

Rather than selling cars to owners, Waymo sees its future in robotaxis. It's more profitable to spread the extortionate cost of the self-driving equipment over cars that are used more intensively. After brief flirtations with creating its own vehicles, Waymo now doesn't bother building cars. It leaves that up to the likes of Jaguar, from whom it's getting 20,000 I-Paces to be delivered from 2020 onwards, and Chrysler, who is in process of supplying 62,000 Pacifica Minivans. Effectively they're treating car companies as suppliers whose kit they add value to through their self-driving conversion.

Analysts rate Waymo as worth between $175 billion and $250 billion – more than Ford, GM, Fiat-Chrysler, Honda and Tesla combined – and think that the major car companies will all be asking it for help in years to come as they strive to make their products autonomous. But traditional car companies are also fighting back by partnering up like never before, both between themselves and with technology companies. They're also madly acquiring start-ups with particular self-driving expertise. The sector has reached fever pitch in terms of new entrants, acquisitions and deals.

Number two in the global self-driving stakes is General Motors. With a distance between engagements of 5,205 miles in 2018, it's some way behind Waymo in the DMV statistics. During that year it travelled nearly half a million miles autonomously in California, meaning that Waymo and GM together accounted for 86 per cent of the autonomous miles completed in the state. GM is a serious contender thanks to, in this case, purchase rather than partnership. It bought 2013 start-up Cruise Automation in 2016 for a reported $1 billion.

Like Waymo, General Motors is concentrating on robotaxis rather than individually owned or leased cars. It won't sell as many but, like Waymo, it thinks it'll make more money out of them than conventional cars. At present a car company earns only one cent from every mile the car travels. When it's in charge of a robotaxi it's earning over twenty times as much. Little wonder that the self-driving division is responsible for a third of the current stock-market value of GM. Unlike Waymo, GM has a factory ready to make all the Chevy Bolt taxis required, which are undergoing tests mainly in San Francisco with additional work in Arizona and Michigan.

Among other deals, Volkswagen has teamed up with Ford who in turn is investing $1 billion in a software company called Argo AI; Daimler is working with a range of tech companies, including Nvidia; and former bankrupt GM parts outfit Delphi has become another major player in the field, buying software makers and driverless car company NuTonomy. The tech giants in China are also muscling in on the action.

Pure tech companies getting a foothold include Apple, who started testing its self-driving cars on public roads in 2017, again in California. It's involvement in the area has naturally encouraged wags to envision their cars made out of aluminium with rounded corners, with proprietary charging plugs no one else uses, non-replaceable batteries and using only the company's own roads unless you buy an expensive dongle. Oh, and obviously they'll get thinner every year. Groan.

The reality is that Apple seems to be lagging behind the competition, with its cars requiring intervention every 1.15 miles, though the company claims it uses a different measurement technique and the figures aren't directly comparable. In 2019 Apple bought drive.ai, a struggling start-up developing self-driving artificial intelligence that could potentially be fitted to any vehicle.

Uber is one company that ranks highly in terms of the public's perception of self-driving cars but badly in terms of

China

China's President Xi is super-keen to make artificial intelligence a vital component of the country's economic roadmap. To that end both start-ups and established Chinese tech giants are part of the action. Despite the ever-escalating trade war between China and the US, it's a process that seems to span the two countries with close links and interdependencies between them. Offices in Silicon Valley send skills back to China, where until recently self-driving car tests on public roads were banned.

Now tests in China are gathering pace. In a process that mirrors the Google-Waymo dominance, Baidu, the operator of China's largest online search engine, completed the country's first autonomous-driving road test based on a 5G-network environment, in Beijing on 22 March 2018, using open-source, autonomous-control software called Apollo. Online retailer Alibaba and tech company Tencent are also developing platforms for self-driving vehicles while partnering with vehicle manufacturers.

Among vehicle manufacturers, Shanghai-based Nio is developing a self-driving car called Eve, while Byton is developing its autonomous K-Byte model with self-driving software provided by Aurora, a company started by the former head of Google's autonomous car programme. Most Chinese companies are aiming only at Level 4 rather than Level 5 autonomy, which may reveal where the world's biggest car market thinks self-driving technology is ultimately heading.

the intervention rate. Human-driver assistance was required every 0.35 miles in 2018. Maybe the extra $1 billion investment the floated company has secured from Toyota and Saudi Arabia will help improve things as it strives to develop an autonomous taxi fleet on land and maybe even in the air.

However, some question if miles travelled between interventions is the right measure of success. Professor Paul Jennings of Warwick University suggests that complex cases are best addressed in simulators or through augmented reality. 'You could condense lots and lots of edge cases in a simulator and get a much better idea of the competence of your system,' he says. With a simulator, a system could be more easily exposed to rare and difficult driving conditions like dazzle from sunlight and water on flooded roads.

One company that does use extensive simulator testing, but doesn't figure in the DMV rankings at all, is Tesla. Tesla's Autopilot system was first fitted to the Model S in 2014, though so far it's really been a driver aid rather than a full self-driving system, offering lane guidance, adaptive cruise control, self-parking and the ability to automatically change lanes and send the car to and recall it from a garage or parking spot.

However, the company has long expressed its intention to offer full self-driving at a future time, acknowledging that legal, regulatory and technical hurdles must be overcome first. Indeed, many owners paid in advance to be supplied with this upgrade when it becomes available. Unsurprisingly, there has been some discontent among those who suspect that day may never come.

Tesla has a different approach to other outfits in the game

in that its systems are almost totally reliant upon software and cameras. Elon Musk doesn't include LiDAR on any of his cars; he thinks it's too expensive and bulky. He also notes that human beings and animals navigate using their eyes, which are the biological equivalent of cameras. According to him, given low prices and ever increasing resolution, optical cameras are sufficient when supplemented by sufficient AI (plus radar and ultrasonic sensors).

'Eyes are basically just cameras. All creatures on Earth navigate with cameras. A fish eagle can see a fish from far away and take into account the refractive index of the water, dive down and get the fish from far away. There's no question that with image-recognition neural nets and cameras, you can be superhuman at driving with just cameras,' Musk said.

The company is accumulating its AI in the form of vast quantities of data about actual driving behaviour. As a Tesla user you're given the option of operating in 'shadow mode'. When you're in it the car isn't taking any action but registers when it would have taken action. If the car is in an accident the company can see if the autonomous mode would have prevented it. The car is also uploading data from its cameras.

Information recorded by your car's sensors gets uploaded to the company and they use it to help make better self-driving AI in the future. Tesla owners report that their cars can regularly upload over 100MB of data a day. This sounds much less than Google's cars but it's often geared to particular requests. When Tesla identifies something it needs more information on, it can request that the cars look out for it and upload relevant data. It can request images of, say, bikes mounted on cars if it is thought that they might confuse the autopilot, so that they can train the system on extra data. The paths taken by human drivers in various situations can be watched in order to analyse what cars are doing before a particular type of event and see whether patterns can be perceived. Unlike virtually every other company in the field, Tesla is keen to offer self-driving abilities on real cars you and I could buy or lease.

How far have we got to go?

On average, a vehicle in the UK would have to travel about 200 million miles before being involved in a fatal accident. That gives some idea of the scale of the safety proofing required by autonomous systems. And even if they do match human drivers, that level will still be way short of what is needed to ensure widespread acceptance.

Some researchers have attempted to do the maths. Tasha Keeney of ARK Investment has suggested that a self-driving car will need to offer 'expected failures' – where the car comes to a halt because it recognises that it can't proceed and signals to the human driver to take over – no more than once every 50,000 miles. That's about how often people break down in the US, apparently, though my personal experience of break-downs is much better. 'Unexpected failures', meanwhile – where the system doesn't recognise something is going wrong and continues to drive on without any signal for help – will apparently be tolerated every 240,000 miles. That's about how often drivers in the US have a scrape of any severity in their cars. It all seems a bit optimistic to me. I think the public and legislators will demand higher standards.

Even Google's own people are cautious. Too many surprising things happen in the driving environment for deep learning to perform well when left unsupervised. DeepMind founder Demis Hassabis has cautioned against self-driving car trials on public roads; he thinks they need to get better in private first. Certainly Waymo has encountered multiple issues with trials of its self-driving taxi service in Phoenix, despite being limited to areas of the city that are almost ideal for self-driving.

An Uber-beater? A Waymo robotaxi pounds the streets of Chandler, Arizona, where the company has launched the world's first commercial driverless taxi service.

The cars tend to stop longer at intersections than human drivers, stop suddenly without warning, have trouble merging into heavy traffic, stick too literally to traffic laws and even fail to recognise red and green lights. According to a report in *The Information*, they're consequently hated by local residents. Playing chicken with self-driving cars has surfaced as a sport for the naughty (and not just teenagers). And Waymo's cars are already the focus of callous recreation. One driver from Chandler, Arizona, admitted hating the cars so much that he tried to run them off the road in his Jeep Wrangler. In one incident he drove head on into a robotaxi that just stopped in time. He said he 'finds it entertaining to brake hard' in front of them. His wife told the *New York Times* that it all started when their young son was almost hit by one of the Waymo cars. It could be a new form of road rage. People have even been witnessed throwing rocks at Waymos, slashing their tyres when they're stopped and even pointing their guns at them and their occupants.

Pedestrians' interaction with normal cars is already to some extent a bravery contest. Each crossing involves a pedestrian calculating the risks between crossing the road as quickly as possible and the chance of being hit. Drivers also have to decide whether to slow down and let the pedestrian cross or keep driving on. In heavy traffic a driver is more likely to yield. Different cultures have different behaviours: compare central Rome versus an English village, for example.

One solution might be to have cars that communicate with other road users. To this end drive.ai displayed messages on its van taxis. For instance, if one of the company's cars stopped for a pedestrian on a crossing, the message 'Waiting for You' flashed up on a screen, along with a graphic of a person crossing the road. Jaguar has shown a car with eyes that interact with pedestrians to show they've been spotted. Cars may need a whole new system of commonly understood lights and gestures just like the ones we already recognise, such as bright red lights for braking and flashing orange ones for the intention to make a turn. How about purple ones for dithering self-driving cars that aren't quite sure what to do next? Judging by the Phoenix experience, though, the cars may need something more radical. At the very least, legislation will need to be tightened so that the cars are not bullied in their moments of indecision.

It should by now be obvious that full automation may still be decades away. However, Level 4 automation, where cars are operating in a well-defined, controlled environment, seems much more achievable. They'll have more in common with driverless public transport than conventional cars. For example, low-speed pods already marketed by companies like Coventry-based Aurrigo and the French company Navya operate as shuttle buses at airports and in retirement communities and universities, with customers sometimes hailing them by app. They're also proving popular in city centres for the 'last mile' of transport to a destination. Navya's latest Autonom cab is already being used in trials in Las Vegas, at the University of Michigan, and in over a dozen other locations round the globe. Unlike Waymo, Navya does not operate the vehicles itself but sells them to other service providers; the current cost of an

Autonom is €250,000. Shuttles like these may well get people accustomed to self-driving vehicles. In the near future, you're much more likely to encounter one of these than a fully fledged autonomous car.

The connected car

Fully developed robo-rides may remain on a distant horizon, but cars are going to become much smarter in other ways too. Using Wi-Fi and 5G they'll start talking to each other more and to the surrounding infrastructure. We're already used to cars being connected to the internet. This is usually achieved via the infotainment systems and our smartphones. Technologies like Apple CarPlay, Android Auto and manufacturers' own systems give real time navigation and internet-based media. Apps can tell us remotely whether an electric car is charged and even switch on the heating.

Vehicle-to-vehicle (V2V) technology goes further. It lets cars broadcast their position, speed, steering-wheel position, braking status and other data to vehicles within a few hundred meters. One use of this technology is platooning. This is where cars or, more typically, trucks link together into a sort of road train. Thanks to communication between the vehicles, some in the train can be driverless – rather like a railway carriage. Drivers in all but the lead vehicle do not need to pay attention to the road, and the vehicles can drive closer together. Because electronic communication is near-instant, they don't need the normal reaction time to operate brakes and they occupy less road space. This, in turn, achieves better fuel economy thanks to less air resistance. Platooning technology could conceivably be retrofitted to existing vehicles, though lead drivers might need additional licences because of the responsibilities involved.

It's feasible already, and companies like Mercedes and Volvo have demonstrated working platoons of trucks. It's even been shown that vehicles can join and leave a platoon with ease. Trials will, however, be necessary to iron out problems and ensure public acceptance. For example, could other cars be prevented from joining or leaving a motorway if there's a long chain of platooning vehicles effectively blocking the exit?

Another potential use of V2V technology is in transmitting information to help others avoid a crash – maybe via a seat vibration or, one imagines, a loud vocal alarm. The cars use the pooled information to build up a comprehensive picture of the events taking place round them, revealing potential dangers that even the most assiduous and alert driver – or the very best sensor systems – would fail to detect or anticipate. Car-to-car communication would also help solve the problem that sensors can't generally see through obstacles.

In live demonstrations on test tracks, cars are shown slowing down to avoid cars out of direct sight that go on to drive through red lights. Similarly cars can adapt their speed to avoid having to stop at traffic lights. Whether they'll easily be able to achieve the same dexterity in more complex scenarios is debatable.

The US National Highway Traffic Safety Administration and the University of Michigan have done a great deal to

The truck in front

Fully self-driving cars continue to prove elusive but the truck without a driver is already here. In February 2018, one designed by Starsky Robotics completed a seven-mile journey at 25 mph on public roads without a single human in the cab. In June 2019, one of its Volvo articulated eighteen-wheelers travelled 9.4 miles along Florida's Turnpike, successfully navigating a rest area, merging on to the highway, changing lanes and maintaining a speed of 55 mph. The secret of these trucks' success is that they can be controlled remotely from HQ in Jacksonville. Though 85 per cent of the trip is done fully driverless, difficult manoeuvres are performed by human drivers sat in an office with a steering wheel looking at screens showing feeds from the truck's six cameras, rather like a video game but for real.

In the US as in much of the developed world there's a huge shortage of people willing to be truck drivers and there's a great incentive to make remote-control trucking work. While there are many other start-ups in the field, it looks like Starsky is ahead for the moment. This truck system would work for driverless taxis only if you were happy to pay a premium to be piloted around by someone sat at a games console somewhere else.

show how valuable car-to-car communication could be in the town of Ann Arbor, Michigan. Between 2012 and 2014, they equipped nearly 3,000 cars with experimental transmitters. After studying communication records for the cars, NHTSA researchers concluded that the technology could prevent more than half a million accidents and more than a thousand fatalities in the United States every year. John Maddox, a program director at the University of Michigan's Transportation Research Institute, reckons that on its own this technology has the potential to transform the way we drive.

The technology could, for example, route information between vehicles and bridges, toll booths, construction signs and other roadside infrastructure. There could even be special lanes where autonomous vehicles are allowed to travel really fast because they're able to share information about incoming and outgoing traffic at great speed, and road signs may become unnecessary because they're fully digitised. On the negative side, though, we may end up with 5G masts every 300 or 400 metres along major roads and only slightly fewer on rural routes. The technology is viable only with a basic minimum infrastructure and a lot of roadworks. More devastatingly, it offers the authorities a previously undreamed of scope for interfering with our driving, in terms of continuously monitoring speed and lane discipline and dishing out penalties and pay-as-you go driving charges in the process. Drivers' costs may spiral with monthly statements, like deranged telephone bills. In the event of a crash your car will automatically shop you to your insurance company.

Brain-to-vehicle technology, or B2V, is an even more ambitious endeavour. It uses electroencephalography to pick up the driver's brainwaves using a cap on the head. I first saw such a device at the technology show CeBit in Hamburg around 2006. I was impressed that someone could use thought to alter a computer mouse and (rather laboriously) type out a message. Since then, it's been refined (a bit) and is used in video games and modern prosthetic limbs.

Nissan has been pioneering B2V use in cars, where it could potentially allow the vehicle to detect your mood as you get in and adjust the climate and entertainment systems to suit. More usefully, the company claims, it will be able to read the driver's reactions to what's happening on the road ahead and translate them into the required action fractions of a second sooner, helping to make driving more relaxing and to avoid accidents. Say the system detects that the driver is about to turn the wheel. Driver-assist systems in the car can begin the action more quickly, maybe 0.2 to 0.5 seconds faster than the driver on their own. Brain-to-vehicle technology will also work to detect when the driver is about to suffer from a bout of road rage and instruct the car's autonomous controls to take over. No more red mist in all its unsavoury permutations.

Given the way we react against technologies that require headgear (think of the failure of 3D glasses and Google Glass), I suspect the equipment will have to be a great deal less obtrusive than the electrode-encrusted shower caps that currently feature in tests. It's safe to say it's in an early stage of development; Nissan estimates it's at least five to ten years away, but as usual

Will 5G really make a difference?

Given the flaky quality of many mobile-phone conversations, you could be forgiven for thinking that the last thing you'd put your faith in for anything safety critical, like driving, is a mobile phone network. But 5G could make a huge contribution to the development of autonomous and connected cars.

It's much faster. All networks suffer from latency – the time it takes for a command to be translated into action. With 3G it's typically 200 milliseconds; with 4G it's around 100 milliseconds. But with 5G it's only 5 milliseconds. So 5G devices, whether they're cars, traffic signs or even bicycles, can communicate with each other in an instant. A driver left to their own devices typically takes between 250 and 500 milliseconds to react.

It can support a far greater number of devices – a million in a square kilometre. With 4G it's nearer 60,000. What's more, those devices can be travelling at up to about 220 mph. In one recent test a McLaren was going at 160 mph round a track and retained a rock-solid 5G connection with a data rate of up to 20 gigabits a second. Breathtaking performance figures.

However, it doesn't always need a network connection: though not ideal, 5G does support devices talking to each other without one.

this is probably wildly ambitious. The safety implications of malfunctions are frightening and there will need to be measures to ensure that other body parts don't generate interfering signals. However, the driver's brain could eventually be recruited to form yet another part of the intelligent car.

What does it all mean?

When autonomous cars eventually arrive, one of our most visible and widely used technologies will undergo a huge transformation. As a result, there is endless speculation about its consequences. The biggest positive, of course, will be a reduction in the number of accidents on the road. Stephen Hamilton, a Cambridge lawyer with a specialty in autonomous cars, suggests 99.7 per cent of road casualties could be avoided in the future. Millions of lives will be saved and millions more people will escape injury. This will lead to a reduction in the amount of healthcare resources required to deal with car crashes and in police time devoted to traffic.

Cars will be lighter and more fuel-efficient if they don't have to withstand crashing, as there will be no need for heavy, energy-absorbing safety structures. Wearing a seat belt will be a practice strictly reserved for the eccentrically pessimistic. Insurance costs will decline rapidly. Autonomy should help efficiency in other ways too. Most autonomous vehicles will not be owned but instead rented out to drivers by the hour; cars therefore won't spend so much time idle and won't spend so much time parked. Parking is a dominant land use in many cities, and self-driving cars might even scoot out of densely populated areas on the relatively rare occasions they're not being used.

Dieter Helm of Oxford University thinks the 'density of use' in cities will increase significantly and quickly. Multistorey car parks could be redeveloped into housing. Even domestic garages and driveways could be turned to alternative uses: expect a lot more man caves across the country. City and suburban streets will also be reclaimed. There will no longer be a need for half their width to be devoted to almost permanently parked cars. Fewer road signs and less road paint will make the cityscape more attractive. Cyclists and other road users will feel safer, and this could encourage further reliance on healthier forms of transport. For the increasing number of people who really are too old to drive regularly (some of them much younger than the famously tenacious Prince Philip), there will be a whole new freedom. The same is true of children and teenagers (and their long-suffering parents, who will no longer have to ferry them everywhere).

An autonomous driver will use energy more efficiently than a human one and won't show a heavy right foot. All these efficiencies will slash the cost of getting about. In March 2017, Barclays estimated that the total costs of car travel will fall from between $1 and $1.60 a mile to just 8 cents thanks to autonomy. Self-driving cars should also provide more efficient links to public transport networks, filling in the blank spots between stations. Out-of-town pubs and restaurants could see an increase in trade as there's no need for drivers to remain sober or for groups to have designated drivers.

Traditional petrolheads may even begin to love self-driving. It could be an enthusiast's dream. Pop the Caterham or the Aston into the pod and get transported to the circuit, or the countryside driving park with its perfect bends and no speed limit; get rid of the boring journey in between, and spend the time on the way being entertained or doing something useful rather than getting frustrated by traffic jams and speed-limited tedium. It's a win all round.

On the negative side, the biggest issue might prove to be the uncertain future facing people who drive for a living. By some measures driving jobs are the biggest sector of the global employment market. Others argue that the efficiency claims are exaggerated. Professor David Begg, an expert in sustainable transport, thinks that cars take up too much space in relation to the number of people they carry, whether they're self-driving or not. Just imagine everyone exiting from an underground station in their own self-driving car – it would be chaos. Central London, for example, already seems packed with fleets of Toyota Priuses thanks to the Uber boom.

More comfortable private journeys with less stress could also prompt people to commute from further afield, generating still more congestion. Might health and fitness actually decline as it becomes easier to make short journeys by hopping in a shared car rather than walking or cycling? And all these cars driving themselves in circles between tasks is in itself a pollution issue.

The technology could also increase a sense of inequality, as it won't be rolled out everywhere at once but instead in limited 'urban islands'. There could be wealthy, safe, automated areas enjoying the benefits of efficient, safe and clean cars, while poorer, more dangerous zones still rely on unpredictable human-controlled cars.

Then there's the housing market and house design. Could your car be transformed into an additional room of your house, or a sort of personal roaming hotel room that ferries you to work? If houses and cars combine, what parts of a country like the UK will be devoted to them and what areas will be saved from the ensuing environmental blight? Areas with industrial parks may become vast parking lots for self-driving mobile homes.

Whatever happens, it isn't going to happen as quickly as the most excited pundits suggest. Self-driving aspirations will continue to collapse and artificial intelligence could well prove unable to cope with the randomness of the open road, with its unpredictable variety and bustle. Cars piloted by drivers are likely to have a fairly long future ahead, albeit one assisted by the connected car's safety-conscious driving aids. I wouldn't normally be so presumptuous as to dole out financial advice but on this I'm fairly certain – if anyone suggests you invest in an exciting autonomous car start-up I'd steer well clear.

TWO

SPARKING INNOVATION

Alternative Power and the Future of the Internal Combustion Engine

At the start of the twenty-first century it looked like the car would continue to be powered the way it had been for the previous 100 years: namely, by oil and the internal combustion engine. In spite of decades of rumours to the contrary, we had plenty of the black stuff and extra reserves were being found all the time. Thanks to a combination of legislation and improvements in vehicles, emissions were getting better too. Likewise, helped by the phase-out of coal for industry and domestic heating, at least in mature economies, our air was getting steadily cleaner. Good news all round.

Even global-warming worries were being calmed by the increasing use of diesel engines. By offering a 16 per cent improvement in efficiency on average over their petrol counterparts they were helping the industry to meet new CO_2 goals. And the assumption was that, by dint of technical advances, the traditional diesel problems of dangerous particulates and nitrogen oxides would soon be things of the past.

But then some developments came along to disrupt this comfortable status quo. Wannabe Bond-villain Elon Musk seized on the fact that lithium-ion batteries – which powered laptops, phones and other gadgets – now made electric power feasible for mass-market cars. Nissan made the first tentative forays into the market with the Altra, launched in Los Angeles in December 1997. But, despite a claimed 140-mile range, the batteries had a limited shelf life and the cars sank without trace. Musk's zeal was something altogether different. He was hugely influential in advancing the idea that the battery-electric vehicle (BEV) might become a credible alternative to those fuelled by petrol.

The original 2008 Tesla Roadster was far from perfect but its perky performance, dynamic handling and lithium-ion batteries combined to smash our preconceptions about electric cars.

Musk believed that tackling the market step by step could transform a notoriously conservative industry. First, he made an electric car on a tiny scale for practice. This was the sporty Lotus Elise-based Roadster, which deliberately went against the traditional economical, box-like, electric-car stereotype and demonstrated that electric cars could be a hoot to drive too. Then he made a luxury saloon to take advantage of the electric car's innate speed and disguise its high cost. After that, he expanded into other areas of the market first by building an SUV, and then by creating a lower-priced, compact executive car. Even better, he made sure his new customers had access to the required electricity. Tesla cunningly provided a network of its own high-speed superchargers in key locations. Initially these were even free for all customers to use.

At the same time as Tesla was taking an industry by storm, health organisations were becoming more vocal in highlighting links between pollution and death rates. Premature deaths and disease were being increasingly linked with pollutants like particulates and nitrogen oxides. Traditional cars and commercial vehicles were seen as a significant part of the malaise. The World Health Organisation declared in 2017 that 92 per cent of the world population and 96 per cent of those living in cities are breathing 'unsafe' air and set a series of target limits – for example, 10 micrograms per cubic metre for particulates smaller than 2.5 microns. Such targets focus minds, especially when they are routinely missed. In 2016, Beijing had an average of 85 micrograms and Delhi a shocking 122. London and Los Angeles were dramatically cleaner, though still above the limit.

Diesels also turned out to be much filthier than promised. A combination of lax testing regimes, the widespread gaming of emissions tests and manufacturer dishonesty exploded with the 2015 Volkswagen Dieselgate scandal. This, along with air-pollution figures and the threat of fines and legal action, prompted politicians worldwide to announce low-emission zones and propose bans on many diesel-engine cars. Madrid, Paris, Athens and Mexico City all announced that they were going to ban all diesels by 2025, while many German cities have already banned older, more polluting diesels.

The threat of global warming meant that the pressures on the industry weren't (and aren't) going away either. In fact, with the new shift away from diesel to petrol, they were increasing, and targets like the EU's promised 66 g/km of CO_2 were becoming harder to meet. Equivalent targets in China and the US were also becoming more stringent. In the US, California and eight other states have a zero-emissions mandate; car manufacturers must sell a certain number of zero-emission vehicles in order to be allowed to sell the rest of their cars. In short, if you're serious about making and selling cars today, CO_2 emissions cannot be ignored.

Anti-internal-combustion-engine sentiment has spread through world politics like wildfire. Norway claimed it was going to ban the sale of new internal-combustion-engine cars as soon as 2025. The UK opted for 2035 and France for 2040. Such aspirations aren't necessarily unrealistic. There's abundant energy landing on the earth every day (solar eclipses permitting). It's just a matter of harnessing, storing and converting it in a way that doesn't cause local and global pollution. There have long been alternatives to mainstream power. But the cheap dominance of the internal combustion engine stopped them taking off. The question is whether the proposed replacements are ready for the big time.

Electric dreams (and nightmares)

The obvious route to salvation is to follow in Musk's footsteps. A new assumption has rapidly emerged that future cars will increasingly be powered by batteries alone. Governments effectively said to car manufacturers: if Tesla can do it why can't you? And an aura of evangelism of the kind you see at Apple product launches started to surround the industry.

As Ian Foley of Equipmake, who designs and manufactures high-performance electric motors, observes: 'If you look at what Musk has done it's very hard to knock the guy because he has almost single-handedly transformed an industry. I was a huge sceptic. I think the tipping point was when he announced the Model 3 and raked in more than $300 million of cash deposits in a week when no one had seen the car. That's what shocked the car industry and made them say, "Crikey, we'd better do something here."'

But there are also hurdles to overcome. These are the same failings that have traditionally limited the appeal of battery vehicles: limited range, slow charging, heavy weight and high cost.

Electric cars are more expensive than their petrol equivalents and still nothing like as useful. In 2018 I spent a week or so with a 40 kWh Nissan Leaf and by the time I gave it back I'd dubbed it the Nissan Stress, such was my range anxiety. I just couldn't get where I wanted to go without worrying. I needed significantly more range and, crucially, the means for faster charging. At a normal person's motorway speeds I found I could travel little more than 100 to 110 miles, not the claimed 150 (why *is* the motor industry forever committed to that gap between actual and claimed figures?). The fastest charger it would accept – 50 kilowatts – could charge it in around an hour, reaching about 80 per cent in 40 minutes. So, not hopeless, but any long motorway journey would be a real chore. On even a fast domestic charger, though, it was a new level of slow. Seven hours for a full charge! As transport delays go, that's worse than the longest airport security queue and slowest rail replacement bus service combined.

All too often the roadside chargers were out of order, or operating on a network that I wasn't signed up to. Even worse, they were rarely plentiful enough that you could leave your car on the charger overnight without fear of breaking new charging-etiquette laws. Would other owners be angry about you hogging the thing? I would get to the hotel after a long day, put the car on charge and, *Oh no*... I couldn't have a couple of drinks at the bar because I'd need to shuffle the car round the car park in an hour's time, taking it off the charger and moving it to a standard space. And it's not just a matter of manners. There can be hard cash involved. Tesla threatens you with 'idle fees' if you leave your car hooked up to one of its chargers after your battery is full – up to 70p a minute in the UK at the time of writing.

The batteries are the main villains behind all these drawbacks.

The limits of lithium

The batteries in cars are typically packs of hundreds or thousands of individual cells that look rather like an oversized pack of AA batteries you might pick up from the supermarket. Combined, they might provide a premium electric car with 90 kWh of battery capacity, about enough to propel the car for 200 miles or so on a motorway. To go the same distance with a combustion engine takes roughly 22 litres of diesel or 25 litres of petrol. But, whereas the diesel fuel weighs under 20 kilos – a figure that goes down as you burn it – that battery pack weighs a mighty 700 kilos, or only slightly less than an entire Mk1 Golf. An electric car is therefore likely to be at least 20 per cent heavier than a petrol equivalent. The longer the range, the bigger the battery and the greater the difference in weight and extra energy required to lug it round.

This isn't likely to change dramatically any time soon. Lithium ion was the big revolution in batteries but they're now a long way down their development path. You can work out the theoretical energy density of a perfect lithium-ion battery by considering the electric chemistry of the elements. It's only about double the best batteries today. And you'll only get that

Charging ahead

Alternative batteries that have the potential to make a substantial difference are still years away, but there are labs full of scientists hoping for a breakthrough.

Solid-state batteries, where the liquid electrolyte is replaced by a solid material, are the closest to fruition. The earliest they'll arrive is the mid-to-late 2020s. Dyson and Toyota, among others, are working on them. They should offer greater capacity, faster charging, longer life in terms of charge cycles and the ability to be moulded into the shape of the car body. The solid electrolyte should also be more fire resistant and work over a wider temperature range.

Lithium-air batteries are closer to fuel cells than batteries and promise huge energy density and thousands of miles of range, but experimental designs have tended to get clogged up as the lithium ions combine with carbon dioxide and water vapour in the air, leading to a very limited shelf life.

The glass battery may sound ridiculous but it's the brain-child of John Goodenough, the inventor of lithium-ion batteries, so must be taken seriously. Now in his nineties, Goodenough says they have prototypes made of just salt and sodium with no combustible or rare earth materials. He claims they have an energy density many times that of current lithium-ion batteries, improved cycle life and much shorter charging times. Other scientists are sceptical.

Supercapacitors can absorb and release power extremely quickly and, unlike batteries, there's no theoretical limit to the amount of power they can store for a given weight. The trouble is that their energy density is currently pretty rubbish, at less than a quarter of lithium-ion batteries. Breakthroughs are on the horizon, though. British company Super Dielectrics Ltd, in partnership with the Universities of Bristol and Surrey, is exploring the potential of using polymers like those in disposable contact lenses to create next-generation storage supercapacitors that will match or exceed existing batteries in energy density. The much-hyped super-material graphene, with its huge surface area, also has the potential to form the structure of future game-changing supercapacitors.

if it has 100 per cent efficiency and zero packaging weight – neither of which we'll ever achieve.

Incremental improvements will come from things like better cooling, but there won't be a revolution and alternative technologies are still years away. Wolfgang Ziebart, who oversaw Jaguar's Tesla-rivalling I-Pace program, cautions against believing battery hype. 'You read every month that someone has invented something great and revolutionised everything. What we have experienced over the last ten years is basically an increase in energy capacity of about 5 per cent per year... and a decrease in the price of the cell of about 10 to 12 per cent a year. Knowing now what's in the pipeline for the next five years I think this is going to continue. But I think we will stick to the current battery technology in principle.'

Whatever the battery technology, you still have to get the power to the car and the physics makes replicating the petrol-station experience impossible. You need to put around 3 megawatts of electricity into a car to have that two-minute fill, but a 3-megawatt cable would be impossibly heavy to pick up; getting such power to the station and coping with the heat generated are also near insurmountable obstacles.

Seasoned petrolheads, like me, are inclined to question how green electric vehicles really are anyway. Heavier than equivalent combustion-engine cars, they produce more particulates from their tyres and brakes. Then there's all the CO_2 embedded in the production of their battery packs, whose raw materials come from people working in miserable conditions in unstable countries like the Democratic Republic of the Congo and Bolivia. Politicians could be about to ban environmentally benign small petrol cars but welcome particulate and CO_2 profligate, politically destabilising battery-powered behemoths with open arms.

Recycling is a big issue with batteries. Lithium is only a small part of a battery and traditionally it hasn't been economic to recycle it. The battery tends to be broken down to form a sort of grease and then incinerated to separate the different metals, but this wastes a lot of the lithium. And as if all that wasn't enough there is still a lack of confidence that battery packs will last the life of the car, which is a worry if you're thinking of buying rather than leasing.

Positive thinking

There are obvious advantages to electric power, though. An electric vehicle powered by a pure electric motor is around 85 per cent efficient at turning electricity into motion and could quite easily get to 90 per cent. By contrast, an internal combustion engine will never get beyond 50 per cent efficiency. The best at the moment are probably about 35 to 40 per cent; the worst convert only 25 per cent of the energy in fuel into motion, wasting the rest as heat. A typical petrol car has to dissipate 100 kW of heat from the engine – hence the grilles and inlets that adorn virtually all cars today. This is terribly inefficient.

Battery cars, with no gearbox and fewer moving parts, are simpler to build and to maintain. They're quiet and smooth compared to internal-combustion-engine cars and often dazzlingly fast, especially when accelerating at lower speeds. I was given the chance to drive Tesla's Model S and Model X shortly after launch. Naturally the press office always supply the fastest version and I can vividly recall the eyeball-stretching acceleration in Tesla's aptly named 'Ludicrous Mode'. By the standards of petrol cars this performance comes tantalisingly cheap. At the time of writing, the bog-standard £77,200 Model S does 0–60 in 3.7 seconds. The performance model with the Ludicrous option costs a mere £14,600 more and, at 2.4 seconds for 0–60, is one of the fastest-accelerating cars in the world. Porsche's list price for the 911 GT3 RS is half as much again and it only manages the sprint in 3 seconds.

Renewable energy is becoming more common and cheaper, so much more of our electricity from our national grid is being classified as zero carbon. That's starting to remove one of the traditional objections to electric cars – that you're just transferring the global warming and other pollution from the car to the power station. And that trend will grow: on a windy, sunny summer's day, well over 50 per cent of UK electricity can be generated renewably. By 2030, it's estimated that renewables will comprise over two thirds of the UK's energy. Similar trends hold throughout the world.

All this makes it very hard to compare the overall environmental impact of battery-powered and conventional cars. MIT has said that an electric Tesla Model S P100D saloon produces more carbon dioxide, at 226 g/km, than a small petrol car such as the Mitsubishi Mirage, which is responsible for 192 g/km. But this is an unfair comparison of a very small petrol car with a large battery one, and assumes that the cars are driven in the Midwestern US, where the electricity grid is carbon intensive. Overall, in most of the developed world, the lifetime CO_2 emissions for an EV – including the global mining and manufacturing, and the electricity to run it – are estimated to be roughly half what they are from a conventional car.

Erik Fairbairn is founder and chief executive of Pod Point, one of Britain's leading suppliers of electric-vehicle charging points, with headquarters in Clerkenwell, London. An entrepreneur by inclination, he ran a supercar club for a while, before having an environmental epiphany. They've rolled out over 70,000 chargers since 2009, nearly a fifth of them in electric vehicle-mad Norway. Their offices are a mixture of the traditional and trendy, with conventional desks and what can only be described as an upmarket shed for meetings. Somewhat to my surprise, I found him agreeing with me about some of the shortcomings of battery power.

'The reason that electric vehicles are only between 2 per cent and 3 per cent of the market today is mainly because the ones that are at a sensible price point don't have enough range. Most people are looking for a car that does 200 or 250 miles on a charge. That's when you're beginning to cover 99 per cent of all the activities that people do.'

He thinks that those marginal improvements in battery power and cost will soon add up to a tipping point. Electric cars are already cheaper to make. They're also cheaper to fuel, at around a tenth of the cost per mile. When the battery costs have come down to the point where the whole car is the same price as a petrol equivalent and gives 250 miles, then 'people will seriously start to engage with electric vehicles'. Fairbairn predicts that costs will continue their downward course. Having been the expensive short-range option, the electric car would then become cheaper to buy and run than its petrol counterparts.

At this point depreciation – another electric-car bugbear – would cease to be a problem. At present, electric cars can depreciate heavily, in spite of healthy new demand, because new ones have much better range. You don't want the old car with a miserably sized battery now there's a new one that goes at least half as far again on a single charge. According to Fairbairn, 'battery packs for most reasonably large cars will plateau around the 90 to 100 kWh mark'. Petrol cars might then begin to depreciate faster themselves as they dwindle in number and face imminent bans. There may even be social pressures in the mix. 'Your peer group questions your decision not to go electric. It's like passive smoking. Your friends and colleagues will know that there's a perfectly decent electric equivalent to the internal combustion engine – which doesn't damage the air.'

As for charging, he doesn't think we need to replicate the petrol-station experience. Electricity is everywhere already. We just need to make sure it flows into your car whenever it's parked. For most people this is probably at home so that's where most charging will happen. In the UK, 7 kW power is the maximum – slightly more power than two kettles. Germans, with their three-phase domestic supplies, tend to opt for a higher 11 kW standard. Either way, a 250-mile range car can be charged overnight... just.

Early electric-car adoption will be driven by people with the luxury of garages and off-street parking. In the UK that's about 60 per cent of households, representing 75 per cent of vehicle owners. Fairbairn thinks it's only when electric-car adoption gets to 10 per cent overall that cities might be persuaded to invest in on-street charging points. These will be installed on the principle that there will be facilities to charge your car wherever you can park, and it won't need to be at high speed. At private establishments, the charging rate and capacity will vary according to the 'dwell time'. With hotels, you're there all night so slow charging is fine; but you might be at a supermarket for only an hour, or at a zoo for five hours, so chargers at such places will need to be faster. Owners of shops and attractions will plan accordingly and may try to attract you with the offer of free power.

On longer journeys you will still want a full charge quickly. Suitably fast chargers are more expensive to install and are likely to be much more expensive to use. The hope is that you'll find the ten or fifteen minutes it takes for an expensive fast boost to get you home on the motorway quite acceptable on the rare occasions you have to do it.

The most powerful chargers are likely to top out at 350 kW, the charging rate used by the IONITY network, which was founded by BMW, Daimler, Ford and the Volkswagen Group, with plans to install stations every 75 miles on average on Europe's main highways. A charger this powerful will add about 240 miles to the reserves of a high-performance electric car in less than half an hour. To avoid compatibility nightmares they're using the Combined Charging Standard (CCS), increasingly adopted by virtually all manufacturers. The network's also compatible with the slower maximum charging rates that will still be used on smaller cars.

Such chargers could become the norm in countries where long-distance road travel is more frequent, like the US or Australia. However, even with the prospect of fast chargers the Australian Prime Minister Scott Morrison has expressed doubts about electric cars: namely that they won't have the range to take his country's families to their favourite holiday

Electric dead ends

Charging electric vehicles is a hassle that needs to be resolved and to that end a number of innovative suggestions have been made. Not all of them are practical (or sensible).

Wireless charging has taken the mobile-phone world by storm. But it is relatively inefficient, especially as power levels increase. Slow charging an accurately positioned car might be just feasible at 3 kW, but wireless on-the-move charging is scuppered by the air gap and the need to switch between many cars using a given stretch of road. You simply can't get enough juice into the car, and that's before taking into account the cost of ripping up roads and installing the infrastructure. Scalextric-type slots for private cars are little better; you can transmit the power but it's very restrictive. The whole point of a car is you can go anywhere you want.

Swapping a discharged battery for a fully charged one can take as little as two minutes. It works for forklift trucks, which are used intensively in a confined space. However, it's just too expensive to keep duplicates of a car's most expensive component on standby in a network of swapping stations.

Solar panels seem like a good idea. One of the world's most famous solar-powered car races has been taking place since 1987 between Darwin and Adelaide, right through the heart of Australia. However, cars that manage to complete the 1,864 miles resemble king-size beds of solar panels on wheels with a tiny bubble for the driver's head to poke through. There's now a 'cruiser' class for slightly more practical machines. The five-seat Stella Vie from the Eindhoven University of Technology won it in 2017 at an average speed of 43 mph.

But few places are as sunny as Australia and it's generally more effective to have batteries charged by static solar panels, which can be angled towards the sun. Small solar panels are appearing on car roofs and bonnets but they only supplement more powerful charging methods. Leaving your solar-panelled car in the sun for a day will probably only get you a mile or two's worth of charge.

destinations, and that purchase prices are too high to meet his political opposition's target to increase their uptake to 50 per cent of all new-car sales by 2030. This would, he thinks, spell curtains for the country's gas-guzzling 4x4s and herald 'the end of the weekend'.

It does all sound to me like it will create a huge amount of roadworks. But even if we get to the stage where you can charge your car wherever you park it, will the power infrastructure be able to cope? There were nearly 32 million cars on Britain's roads in 2018. Imagine them all charging at the same time. It would be like half-time tea breaks in the World Cup but thousands of times worse. There's only enough capacity to charge about a million and a half cars if you try to charge them at peak times – like 6.30 on a winter's evening, when people are coming home from work, switching on the heating and cooking supper. But when you consider capacity across the grid over twenty-four hours, the system could probably cope.

Success will be down to the management of power through smart charging systems. Apps will enable charging to be scheduled during the night when there's less demand and when cheaper electricity is an incentive. On shared public systems or in office car parks the charging posts themselves can talk to each other to spread the load. App-controlled overrides would be possible if you needed to get going in a hurry, at an extra cost.

Erik Fairbairn sees battery-car sales becoming the majority by the mid-2020s and rising to 90 per cent of European new-car sales by the end of the decade. Most in the European industry are far more cautious, though, estimating that a third of the market will be plug-in by 2030. Wolfgang Ziebart reckons that battery-electric vehicles will achieve a 10 per cent share in Europe by 2025, up from 1.4 per cent in 2017. He thinks the share will be a little less in the US, but rather more in technologically keen countries like China.

I would bet on the more conservative projections. Even after hearing from a true battery-car enthusiast I couldn't completely shift my feeling that for many drivers there'll still be too much range anxiety and loitering at charging points. Fortunately there are other technologies brewing that will better suit their needs.

Hydrogen

Hydrogen has long been touted as the ideal future power source for cars, because its only by-product is water rather than carbon dioxide or other nasties. While it can be burned directly in an internal combustion engine, that's not the most efficient way of using it. The preferred route is to use a fuel cell. Hydrogen goes into it and reacts with oxygen to produce electricity and water in the presence of a catalyst, typically platinum. The electricity gets stored in batteries (or sometimes supercapacitors) and drives an electric motor as in a battery-electric vehicle. There's no air pollution and, if you can generate the hydrogen using renewable energy, no global warming either. Unlike battery cars, there's no range anxiety. You can refill a hydrogen car nearly as quickly as a petrol one. It sounds like the perfect power source.

It's therefore both surprising and depressing that people who like battery-electric cars often seem to hate hydrogen. Rather like the argument between nuclear power and renewable energy, it generates heated debate among those who largely have similar aims. To me it brings to mind a famous scene in Monty Python's *Life of Brian*. The hapless Brian mistakes the activists of the anti-Roman People's Front of Judea for the activists of the Judean People's Front. Reg, the leader of the People's Front, is not pleased: 'The only people we hate more than the Romans are the f***ing Judean People's Front.'

Elon Musk himself is one such hydrogen sceptic. Here's a typical rant: 'I just think hydrogen fuel cells are extremely silly… It's just very difficult to make hydrogen and store it and use it in a car. Hydrogen is an energy-storage mechanism; it's not a source of energy. You have to get that hydrogen from somewhere. If you get that hydrogen from water… electrolysis is extremely inefficient as an energy process.'

He goes on to say that 'Hydrogen has very low density, it's a pernicious molecule. It likes to get all over the place. You get metal embrittlement from hydrogen, you get hydrogen leaks, it's an invisible gas, you can't even tell that it's leaking; it's extremely flammable when it does and it has an invisible flame. If you're going to pick an energy-storage mechanism, hydrogen is an extremely dumb one to pick.' Doing the rounds on the internet is a diagram that explains what he means. Once you take into account generating the hydrogen from water and storing it, electric vehicles appear much more efficient. With them you're looking at keeping 69 per cent of your original energy at the wheel compared to only 19 per cent with hydrogen.

In addition to issues of efficiency, safety and transport there's the problem of very little existing hydrogen infrastructure. In the UK, for example, there are only nine filling stations and they're concentrated in and around London. On top of this, the fuel cells themselves are incredibly expensive and most hydrogen production in the world currently comes not from renewable electricity and electrolysis but from natural gas, via a process called reforming. This involves exposing the

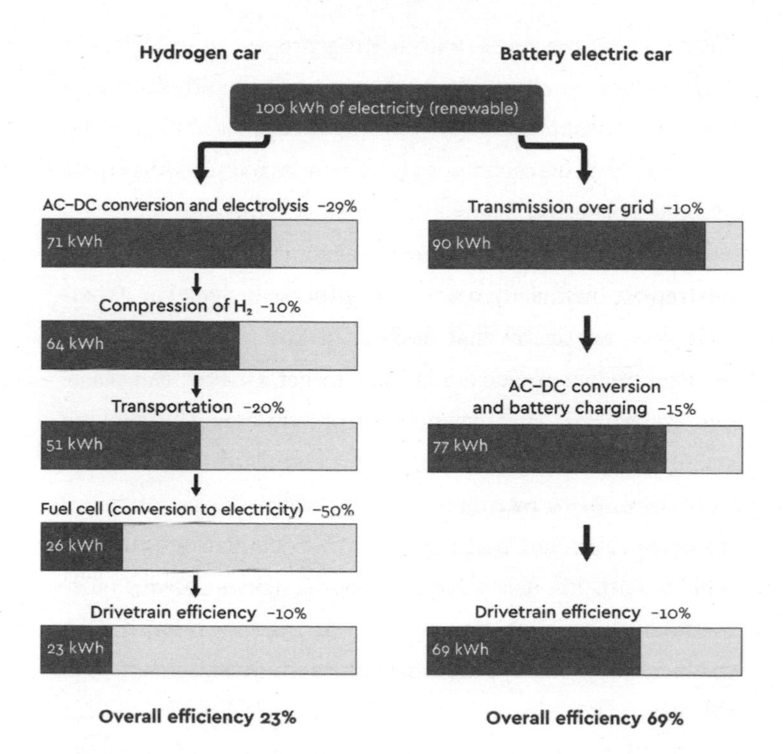

Hydrogen versus batteries – where the energy goes.

gas to high-temperature steam, which causes it to separate into carbon monoxide and hydrogen. The carbon monoxide itself is then reacted with water to produce more hydrogen and carbon dioxide. Obviously the process doesn't do much to help reduce CO_2 emissions.

On the positive side hydrogen is, by weight, the most energy-dense material on the planet, and it's everywhere – in water. Some of the biggest players in the car world think that any difficulty you can name with hydrogen is surmountable and that once you overcome them it has an unrivalled ability to fuel the world's sustainable-transport needs. Toyota, Hyundai and Honda all back the technology and Daimler has now revived its hydrogen plans too.

The car manufacturer that's been most enthusiastic for the longest time is Toyota. It started its hydrogen-car experiments as long ago as 1992 and has been steadily working on eliminating hydrogen's drawbacks as a vehicle fuel. As with its hybrids, the company has taken a very long view. The thinking is that it takes forty years to develop an automotive technology from the drawing board to the mass market. There could be a tortoise-versus-hare battle going on. Musk is getting to market quickly with his flashy batteries but Toyota's slower, more methodical approach, though it may at first appear lumbering, might actually become dominant and achieve better results in the end.

I spoke to Jon Hunt, Toyota's alternative-fuels manager in the UK. He points out that his company was searching for an energy-dense battery long before it made cars, in the days when its primary business was making weaving looms.

'Back in the 1920s, Sakichi Toyoda, who was the grandfather of the company, thought electrification would be the best route for looms and offered a prize of a million yen for whoever could come up with a battery that was the equivalent of petrol in terms of energy density. Nobody's managed to win the prize. The chemistry of batteries means it's impossible to get the energy you get in the equivalent litre of petrol or diesel.'

Though Toyota does build battery vehicles, it's concentrated its efforts on hybrid technology – and its hydrogen fuel cell research started as a spin-off from that. The fuel powers electric motors directly, while batteries or supercapacitors are on hand to store any energy recovered during deceleration and provide extra power while accelerating. In effect Toyota took its hybrid powertrain and mated it with a fuel cell. The first commercial car was the limited production Mirai of 2015. *Mirai* is Japanese for 'future'.

There are various types of hydrogen fuel cells used in cars. Toyota chose a so-called 'proton-exchange membrane' because it's lighter than other types, though it does require precious metals as a catalyst – in this case platinum again. The design was improved by making it virtually maintenance free, apart from the occasional filter replacement and water top-up. The company is gradually reducing the amount of platinum required – it's now only marginally more than in many conventional cars' exhaust systems and it's also completely recyclable. Newer cells used by Mercedes and Hyundai have also further reduced platinum usage and there's a lot of work going on into using alternative, non-precious metals as well.

The biggest challenge with hydrogen isn't really the fuel cell, though. It's the fuel storage. Hydrogen has a curious property whereby it can diffuse into metal structures and cause all kinds of problems. The metal becomes more brittle, which means that conventional fuel tanks would need to be far too heavy and costly. They'd also need frequent replacement as the compression and decompression of hydrogen during filling and emptying causes expansion and contraction of the metal.

Toyota's solution was to develop a carbon-fibre tank with an inert glass liner that can contract and expand without any fatigue. It is light and cheap to make in volume.

With safety concerns about hydrogen refuelling, Toyota then helped develop an international standard to refill cars with a universal high-pressure gas nozzle and a standardised connector. It's a contrast to the bewildering variety of plugs and connectors that complicate life for battery-car users. The fuel itself is also standard, unlike electricity with its different voltages and different outputs.

When I drove a Mirai I found it surprisingly normal. It feels like a very competent electric car. It's quiet, accelerates smoothly and feels brisker than its 0–60 time of 10.5 seconds would suggest. The fuel cell takes up some luggage and passenger space but it's still reasonably roomy and comfortable for four. The range is 300 miles and the cost per tank is comparable with petrol. At the end of a journey a tiny splash of clean water dribbles out of a hole under the boot floor – this is its only emission.

At present the price is a not insignificant £66,000. Jon Hunt anticipates that that will fall by 20–25 per cent per generation, as with hybrids. The drivetrain's the same as in a current hybrid, so a similar cost already. The costs of the fuel cell and tank will fall too as they enter mass production in Japan – from its current cottage-scale 3,000 a year to the tens of thousands. 'I think the third-generation car will probably be mass market in pricing,' says Hunt; it is slated to arrive around the late 2020s.

Public hydrogen-fuelling stations currently cost about four times as much to build as their petrol and diesel equivalents – another considerable obstacle to widespread adoption. But even a skeletal network would be a start. The UK H_2 mobility team at Toyota, tasked with investigating just such things, estimates that you can get away with just sixty-five stations on the main arterial routes, but that for greater convenience you need a maximum of twenty-five to thirty minutes' journey time between stations, which equates to a figure of 1,100. That's still much less than the 8,000 or so fuel stations the UK has currently. Building such a network will take lots of time and money, but it will need to happen if a solution is to be found for the chicken-and-egg problem that always seems to plague discussions about hydrogen. Why would you build the filling stations if there are no cars, and why would you buy a car if there are no stations?

The incentive to make this massive investment increases when you look at hydrogen's potential to provide fuel right across the economy. This is a fuel that can power industry, homes and transport in all its modes. The Hydrogen Council, initiated in Davos in 2017, is a global group of CEOs committed to developing a hydrogen society. Representing over thirty companies, they've come to the realisation that it's not sustainable to maintain our current energy distribution and production. When you look at industrial, transport and domestic uses there's a synergy you can achieve with hydrogen that you can't with anything else.

Heavy transport would make the transition first because the

weight problems of batteries are vastly more severe with heavy trucks and long journeys. You need tonnes of batteries, days of recharging – and all those batteries really eat into your payload. In the US, Coca-Cola, Walmart, Amazon and others have made significant investment in fuel-cell forklift systems that use hydrogen generated on site, leading towards a sustainable zero-emission energy-production system that will power their factories and take them completely off grid. Giant US brewing outfit Anheuser Busch is going further and looks set to deploy huge hydrogen-powered interstate haulage trucks, made by a company called Nikola.

Nikola is building a network of 700 fuelling stations in the US and the order book for its trucks grows by the day. Some London buses and German trains already run on hydrogen fuel-cell power – further examples of how commercial vehicles are forging the foundations of hydrogen-powered transport.

As for worries about efficiency, in many parts of the world the percentages highlighted by Elon Musk aren't really an issue. In Orkney, just off the mainland UK, the prodigious tidal and wind energy infrastructure frequently generates more electricity than the local grid can accept. There's now an electrolysis machine to use some of the surplus to convert water into hydrogen that's then used as a substitute for fossil fuels. Elsewhere, there's a strong economic case for putting free, surplus energy to use in the production of hydrogen that can then be sold on. As more national grids are powered by renewable energy, efficiency will become less relevant.

One radical hydrogen car coming to market is the Rasa. Surprisingly, this revolutionary vehicle isn't being manufactured in Stuttgart, Tokyo or even California. It comes from Llandrindod Wells, Wales. Rather than converting the architecture of a conventional vehicle to hydrogen power, a company called Riversimple, headed by Hugo Spowers, is completely rethinking the car and its ownership experience around the current limitations and opportunities presented by the hydrogen fuel cell.

Spowers seeks to eliminate the environmental impact of personal transport. 'We can have sustainable cars but not if we blunder along blindly just making things a little less bad. Elimination, not reduction, has to be our goal. Being less unsustainable is still not sustainable.'

He was inspired by the work of American physicist Amory Lovins of the Rocky Mountain Institute, who invented the term 'hypercar' in the 1990s – long before it was used to refer

The hydrogen-fuelled Riversimple Rasa is an instructive exercise in reducing the car's environmental impact.

to super-powerful luxury sports cars. Lovins's vision was for a vehicle that has little or no environmental impact; this would be achieved by making it out of ultra-lightweight materials and powering it by a renewable, sustainable and non-polluting source of energy. The Rasa is the closest I've ever seen to anyone making this dream a commercial reality.

Weight is the first priority. Making the car as light as possible means that a lower-powered fuel cell can be used. The cost of lower-power fuel cells is much lower than higher-powered ones because the precious metal plates have a much smaller surface area. (This is unlike petrol engines, where the production costs can be quite similar for different power outputs.) The Rasa's body is made from carbon fibre. At 580 kg it's half the weight of a conventional small hatchback and lighter than the battery packs fitted to some electric cars. It's so light that the hydrogen fuel cell needs to generate only a tenth of the power of the engine in a normal car, around 11 bhp. Although carbon fibre is relatively expensive – the 39 kg of the material in each Rasa costs about £2,000 more than a body made of steel – the benefits in reduced weight more than offset the additional cost in the longer term.

Whereas the hydrogen-powered Toyotas and Hyundais have a weighty bank of batteries charged by the hydrogen fuel cell, and a conventional electric motor, the Rasa has a miniature electric motor in each wheel, which makes the car effectively four-wheel-drive. Most of the Rasa's power is 'live' electricity, generated by the fuel cell, but it also recovers electric power from braking. It does this more efficiently than most battery and hybrid cars. Normal braking power is achieved by running the electric motors in reverse; the electricity generated is then stored in supercapacitors rather than batteries. Because these store energy as electricity, rather than in chemical form, they can hold and release power more quickly, which aids acceleration and efficiency. While a Prius recovers around 10 per cent of the kinetic energy involved in braking, the Rasa recovers 50 per cent. The supercapacitors are also constantly charged by a trickle of electricity from the fuel cell, so there is extra power on hand if needed. Spowers explains his sums:

'In a normal car the engine is sized for maximum acceleration, which you only use for less than 10 per cent of the time. When you're cruising on a motorway a typical car uses only 15 to 20 per cent of its peak power so it means that the engine is 80 per cent redundant for 90 per cent of its life. What we've done is sized our fuel cell for that 20 per cent, not the 100 per cent, and relied on the supercapacitors for the other 80 per cent when you want to accelerate.'

Aerodynamics is also vital. The car has a classic teardrop shape with surprisingly high ground clearance. As a car moves, the air is sheared between the stationary ground and the moving floor of the car. So, the further you can get the two surfaces apart the better. The inspiration here is work done by Dr Alberto Morelli in the 1970s and seen in the Californian-built Aptera. This wingless aircraft was not a practical real-world car, but the Rasa takes some of its elements – the high ground clearance, the use of teardrop and elliptical shapes – to achieve low drag.

In spite of all the weight-saving and technological wizardry, there's no way the Rasa could be sold profitably in a conventional ownership model. Because it is built by a low-volume specialist car manufacturer, it would cost about three times as much as normal cars. Spowers's solution is to sell the car as a service for a specified period, typically one to three years. The package includes maintenance, tyres, fuel and insurance. There's a fixed-price element and a mileage rate.

By looking at costs from manufacture to end of life, Spowers reckons he's in a position to tackle ordinary cars in terms of value for money. Conventional car manufacturers make most of their money by selling the car and a few branded parts in the early years of a car's life, so they need you to buy a new replacement as soon as reasonably possible; this business model does of course encourage built-in obsolescence. With the Rasa, Riversimple pay for all costs and are therefore incentivised to keep their investment going as long as they can. The result is much less wasteful.

Gull-wing doors add a touch of glamour, the suspension's been designed by former motor racing engineers, and 0–60 acceleration takes a respectable 10 seconds. But 60 mph is also the maximum speed, to save energy, which would be rather tiresome on long journeys. Over 2,000 people have expressed an interest in taking part in the initial trial of 20 cars. I was fortunate enough to meet some of them at the launch in Abergavenny. There was a palpable Welsh enthusiasm in the air. Some were keen to swap their second car for a Rasa. Most seemed happy with the rental of £370 a month and an 18p mileage charge to cover fuel and other consumables.

The car is a technological tour de force and the company could well have a future marketing its technology. As proof of concept, I hope the Rasa succeeds; but I fear the car itself may struggle. It's competing in the area of the market where cheap battery cars are most likely to be successful. And overall, I'm still left with the impression that the internal combustion engines won't be comprehensively bested for some time yet – either in terms of practically or enjoyment. But how can they be cleaned up?

Hybrids

Happily for those of us who still enjoy the roar of an internal combustion engine, cars have continued to become massively cleaner in terms of local emissions and they are more efficient too. Furthermore, while they still emit CO_2, they contribute less to global emissions than you might think.

The main way in which the internal combustion is now being goaded to a greener future is by combining it with battery power. Though car manufacturers and politicians are saying they'll eliminate the internal combustion engine from the streets, read the caveats and they're often restricting the cull to new vehicles powered solely by petrol or diesel. The venerable technology will still have a major role in a hybrid system or as an additional motor.

For an increasing number of drivers in the next decade or so the car of choice is going to be a petrol hybrid. Though hybrid

How bad are cars really?

You might think that cars are the main contributor to carbon emissions but you'd be wrong. According to the US Environmental Protection Agency just 6 per cent of greenhouse-gas emissions are down to cars and vans globally, although they contribute more in certain countries (with 19 per cent in the US). Nevertheless, electricity, heating, agriculture and industry are all much bigger villains.

Similarly, some of the limits laid down by the World Health Organization and others may need a reality check. Cooking a Christmas dinner can raise particulate matter in your home to twenty times WHO limits for more than eight hours, according to research carried out by the University of Colorado, Boulder; meanwhile Germany's Fraunhofer Institute has highlighted limit-busting levels of NO_x when you light up your gas hob. Researchers have found particulate pollution in London's tube stations is up to thirty times higher than beside the city's busy roads.

The late Tony Frew, who until recently advised the UK government on the medical effects of air pollution, suggested we balance the risks against the benefits. He agreed that high levels of pollution in places like Beijing or Delhi should be addressed. But he questioned how much political and financial effort should be expended to bring places like London, LA or Nairobi – all of which have much lower levels of pollution – below WHO guidelines.

'If you got rid of all the transport in London you might be able to reduce the particle concentration by about 2 micrograms per cubic metre,' he said. 'At the moment for millennials, median life expectancy is into the mid-nineties. If you get rid of that pollution and nothing else happens you might live [one month longer]. We have to ask ourselves whether that extra month is a worthwhile benefit in return for not having any cars or any public transport or any delivery vehicles.'

Sir David Spiegelhalter of the Winton Centre for Risk and Evidence Communication at the University of Cambridge says: 'There are huge uncertainties surrounding all the measures of impacts of air pollution, with inadequate knowledge replaced by substantial doses of expert judgement. These uncertainties should be better reflected in the public debates.'

technology has been around for years in one form or another – the first diesel-electric trains appeared as early as the 1920s – hybrid technology really got going with the launch of the Toyota Prius in Japan in 1997.

The Prius is still the most popular form of hybrid car, what's commonly termed a 'full' hybrid: in this case, a petrol engine combined with an electric motor, with each being powerful enough to drive the car. A sophisticated control unit distributes the power seamlessly from each power source to the wheels via an automatic transmission. Toyota has sold over 6 million of them and you've probably driven or at least travelled in one. In London, for example, they are ubiquitous as the choice of Uber drivers. Of course, Toyota is not alone in manufacturing full hybrids; they are also produced by Audi, BMW, Citroën, Land Rover, Lexus, Mercedes, Peugeot, Porsche and Volkswagen, amongst others.

Using two engines where one would do may seem obviously inefficient but it does have advantages. The idea is that each engine is used where it is most efficient. Electric power is better when setting off, or in stop–start driving at low speeds, as it exploits the abundant torque (best thought of as the twisting effect) of the electric motor. The petrol engine provides more power and is therefore better suited to relatively high-speed cruising. When maximum power is needed, both engines can be used together – meaning that the petrol engine can be slightly smaller for a given level of performance. Crucially, like battery and hydrogen cars, the full hybrid also harvests some of the energy that would otherwise be lost in braking and uses it to recharge the batteries.

The petrol engine tends to take over at about 25 mph. They operate together only when maximum acceleration is required. (You can see this in a graphic on the dashboard, but in my experience the novelty wears off quickly.) Batteries in most hybrids are smaller than those in full-time electric cars. The range in electric-only operation is usually just a mile or two, as the cars are not designed to be used in this way. They've already started to rival the economy of diesels and should soon be able to offer similar performance – in particular better acceleration – as they become the norm for medium-size everyday cars upwards.

Smaller cars will increasingly be so-called 'mild' hybrids. Here a relatively small conventional engine uses an electric motor, mounted 'in parallel' between the engine and gearbox, to give extra help when required. The big difference is that the electric motor is not capable of powering the car on its own. The batteries regenerate when the car's driving normally and provide small but significant assistance with launching off from rest. Again the idea is that you can have a smaller, more economical petrol motor for a car of a given size, and hence greater economy. Mild hybrids will deliver a 10 or 20 per cent improvement in fuel efficiency for a petrol engine and enable them to more effectively take the place of diesels. Such technology will certainly take over from small, high-revving turbocharged engines, which produce unacceptable levels of nitrogen oxide and aren't anything like as economical in real life as they are in the test lab.

'Plug-in' hybrids are closest to pure electric cars. As well as being boosted by the engine and braking while on the move, these cars can be plugged into the mains to charge the batteries for longer battery-only journeys. The batteries are typically larger than in a full hybrid and in some models can give a range of about 30 or 50 miles of electric driving. Mitsubishi's Outlander plug-in hybrid has proved particularly popular in Europe. Increasingly, on new models, you'll be able to use the engine to charge the batteries as well. So for a long journey between two urban areas you can start off on battery only, switch over to combustion on the motorway and recharge at the same time so your battery is full when you reach your urban destination for no tailpipe-emission travel there.

It's often been hard to justify the extra expense of buying a hybrid because the fuel-consumption benefits are marginal. But combustion-engine cars are necessarily becoming more complicated just to pass tightening emissions tests, so over the course of time the cost differential will likely disappear. Nevertheless, as with many car statistics, hybrids' advertised fuel efficiencies can be misleading. Furthermore, with some of them you can actually end up with worse fuel consumption if you don't plug them in or if you use them almost entirely on motorway journeys – essentially you're running a normal combustion-engine car and just lugging around a heavy payload of batteries.

Take that hugely popular Mitsubishi Outlander. In 2018, it was the UK's most popular plug-in hybrid. Its official fuel-economy figure is a staggering 131 mpg and its official CO_2 emissions are a mere 46 g per 100 km, making it exempt from the London Congestion Charge and subject to tax benefits. But this really is playing the system.

These positive figures are based on a test cycle that starts with a fully charged battery when the Outlander can travel only about 28 miles on battery power alone. Then the petrol powertrain kicks in and the Outlander becomes a large petrol car carrying around a couple of heavy electric motors and a battery. In reality it would be a truly eco-friendly car only if you did short journeys (under 28 miles) and charged the batteries in between. How many people do that?

The emissions expert Nick Molden recognises the hybrid's capacity to be a goodie-two-shoes and a very naughty boy. 'Even to say "hybrid" you have to be careful because it covers a multitude of sins. The Prius is probably the best in class. The fourth generation of Prius is an incredibly clever vehicle. [Toyota] created a 10 per cent fuel-economy jump in one generation. From what I understand, for every turn of the engine they decide if it is more efficient to fire the engine, or use a bit of electricity. The devil's in the detail. What is the exact nature of your hybrid technology? It's becoming like internal combustion engines. You get good ones and bad ones. We should judge hybrids at the model level rather than at the generic "if it's a hybrid it must be good".'

If all this talk of carbon emissions and efficiency is sounding rather dull, don't worry. There is a special form of hybrid that deserves a category all by itself – the hybrid hypercar. The first was the Porsche 918 Spyder. It produced a stupendous 887 horsepower through a combination of a 4.6-litre V8 petrol engine and two electric motors. Porsche quoted a 0–60 time of 2.5 seconds, the price was £624,000 and it could do 214 mph. And yet, according to official figures, when it launched in 2014 it was so economical that it was exempt from the London Congestion Charge. The battery-only top speed was 100 mph.

Others followed at similarly eye-watering prices: the McLaren P1 at £866,000 with 727 bhp, and the Ferrari LaFerrari at £1,150,000 with 950 bhp. The hybrid-hypercar concept is now being taken to further extremes. The 1,035 bhp, £1.75

million McLaren Speedtail has a V8 petrol engine – but a staggering 309 of those bhp will come from its electric motor, even though it can't power the car on its own. The dramatic extended rear bodywork is to promote aerodynamic efficiency at its 250 mph maximum speed. Aston Martin offers the 1,160 bhp, £2.5 million Valkyrie: 1,000 bhp from a naturally aspirated 6.5-litre V12 combined with 160 bhp from an electric motor.

Most ambitiously, the 1,000-plus bhp Mercedes-AMG One aims to bring Formula One hybrid technology to the road. The rear wheels are driven by a turbocharged 1.6-litre petrol V6 with direct injection and 11,000 rpm. There are four electric motors. Two drive the front wheels plus there's a motor directly linked to the crankshaft to boost acceleration and another built into the turbocharger to reduce lag. The engine is consequently highly efficient, but has to be dismantled and rebuilt after about 30,000 miles to check the integrity of the components. The 275 cars being built at a cost of £1.9 million each have all been sold already. Mercedes received over 1,000 orders but have restricted production to keep the car exclusive.

Falling in love again with diesel... maybe for a bit

But do we need hybrids at all? In the short term, the miserly motorist's best friend diesel may well have a comeback. Finally, after years of false promises, the fuel is now much cleaner. It seems that Dieselgate spurred manufacturers on to almost eliminate its issues. As we wait for battery-powered and

Dieselgate

As soon as governments began vehicle-emissions tests, car manufacturers started cheating them. However, the 2015 Volkswagen Dieselgate scandal made previous incidents look like petty crime. VW were in a major push to sell diesels in the US and trumpeting the cars' low emissions, which were miraculously achieved without costly and inconvenient tanks of exhaust treatment liquid.

In reality they'd rigged as many as 11 million diesel cars globally with 'defeat devices' – cheat software that detected when they were being tested under laboratory conditions and reduced emissions accordingly. Out in the real world, the pollution controls were turned down or switched off and NO_x levels soared as high as forty times the legal limit.

When this was discovered by researchers examining the differences between laboratory tests and real-world emissions, the consequences were huge. In the US, VW agreed to buy back or fix 475,000 cars, owners getting compensation of up to $10,000 each. Seven employees have been sentenced, including executive Oliver Schmidt, who was given seven years and a $400,000 fine after admitting that he helped the firm evade clean-air laws. It has cost the company over $30 billion and rising.

More widely the saga exposed widespread gaming of tests by manufacturers. Pollution has been much higher than it should have been in many countries and nobody believes car-industry pollution figures any more. Ironically, the scandal may have been the final push needed to clean up the notoriously dirty fuel.

hydrogen-fuelled cars to become mainstream, diesel cars will help minimise the environmental impact of keeping us moving thanks to their relatively low CO_2 levels that in many cases outperform petrol hybrids altogether.

Making (rather than faking) diesels cleaner has made them more expensive and more complicated. As well as particulate filters the latest diesel-engine models have tanks of exhaust fluid, a colourless, non-toxic mixture of urea and de-ionised water, often termed AdBlue. Tiny amounts of this are injected into the flow of exhaust gases and at high temperature a chemical reaction takes place that turns the harmful nitrogen oxides into harmless nitrogen and water.

It works. Under real driving conditions NOx is down as low as 20 milligrams per km. The regulated level is 80. Even the last generation of vehicles were regularly putting out 800. What's more, Loughborough University claim they've discovered a technological breakthrough that will get diesel NOx down to 13 milligrams per km. Arguably, they don't even need to.

While 20 milligrams isn't zero, we're at the point where there are other, much bigger problems in terms of urban pollution – like wood-burning stoves, construction equipment, domestic heating and cooking, and even the modest candle.

A modern diesel produces less particulate matter than a battery-electric car, which is typically heavier and creates more dust from tyre and brake wear. The particulate filters are now so good that when the car is driving through a polluted urban area it sucks in more particulates than it emits, capturing them in the car. The air coming out of the tailpipe is cleaner than the air going in. Nick Molden calls this 'net negative emissions'. In contrast, the electric vehicle with no tailpipe and no filters has no such cleaning effect. These new cleaner diesels even produce fewer harmful particulates than the latest petrol cars.

The problem is what to do with older, more polluting diesels. In Europe the new standards include on-road testing but apply only to model approvals from 1 September 2017 and to all new registrations from 1 September 2019. The diesel clean-up has only just started.

The previous standard, which applied from 2014, incorporated some good and some very bad cars. The combination of a lax testing regime and manufacturers' willingness to bend the rules meant that in Europe, although no one actually broke the law, some cars were massively over-emitting. These 'Euro 6' diesels currently attract no extra charge in London's Ultra-Low Emission Zone and German cities haven't banned them. But if the air stays dirty they won't stay on the road for long and many people will be incensed. Nick Molden reckons that manufacturers may start feeling obliged to protect their reputations by offering big incentives for owners of new-ish but dirty diesels to swap them for something cleaner.

These newfangled diesels will also need to be carefully maintained. To get these cars so clean involves an array of complex systems. It only requires one bit to break down and the car goes back to being like a cigar smoker in a pub. A more detailed MOT test may need to be introduced. So-called plume chasing might also be proposed. Police patrol vehicles would pull in behind cars and physically detect what comes out of the tailpipe while you're driving; it takes around 30 seconds. They'll be a bit like speed cops but sniffing tailpipes rather than checking radar guns.

Fuels of the future

Other fuels can make internal combustion engines more environmentally friendly too. One ambitious idea is synthetic fuels. These would use CO_2 captured from the air, which would then be combined with hydrogen created by renewable electricity. Giant processing plants would be erected in parts of the world with surplus solar energy that could be used in the conversion. Big fans draw in the ambient air across plastic corrugated sheets coated with potassium hydroxide to extract the CO_2. A typical plant should be able to extract about a million tonnes a year. This is equivalent to the CO_2 produced by about a quarter of a million cars.

Deserts would be ideal locations. Professor Jamie Turner from the Department of Mechanical Engineering at Bath University's Powertrain and Vehicle Research Centre thinks that 770 square miles of the Sahara could supply Europe's entire transportation needs. Although synthetic fuel would release carbon dioxide on combustion it would be a closed loop, because the fuel is derived from CO_2 already in the air.

The principal problem is that the process is very inefficient. Because hydrogen needs to be generated from water and because the carbon is captured from the air, the current overall efficiency is just 15 per cent. Consequently costs per litre could be several times higher than conventional fuels, even when producing at scale.

Wolfgang Ziebart tells me that 'Synthetic fuels will be required where energy density is the most important thing.' He has little doubt that aeroplanes will be using synthetic fuel in the future as it has the highest amount of energy per kilogramme. Mazda is trying to make synthetic fuels biologically in a laboratory, as a way of reducing costs, but has so far succeeded in doing so only at a miniature scale.

Liquefied petroleum gas (LPG) or autogas could also bridge the gap before hydrogen and electric become widespread. The fuel has been around since the late 1920s. It is produced as a by-product of natural-gas processing and is currently used to power just 3 per cent of the world's cars but is very popular in certain countries. In Armenia 20 to 30 per cent of the car fleet runs on it. In Turkey it's 10 per cent. Poland, Italy and Japan are all large markets. It's estimated that here in the UK some 170,000 cars have been converted to run on LPG. And with 1,400 stations selling it the infrastructure is already there, so filling up isn't a problem.

Most petrol engines can be converted and independent tests have shown that CO_2 emissions are 20 per cent less than petrol on average. However there are good reasons why it may never make the mainstream. Cars while they're warming up still run better on unleaded, so the original fuel tank and fuel system need to be retained. A typical car uses a tank of unleaded to every five of autogas. The size of the extra fuel tank encroaches on luggage and conversions aren't particularly cheap at around £2,000. It can take over 30,000 miles of running to get the money back. LPG is also heavier than air and tends to settle in low spots. For this reason it's regarded as an explosion hazard and is banned in many car parks and in the Channel Tunnel – so make sure you take the ferry and let's hope it doesn't go up in flames.

Natural gas also has its drawbacks. It's an energy source that's in great supply at the moment thanks to shale-gas discovery and exploitation. According to NGV Global, who collects statistics on such things, there will be 30 million gas vehicles by 2024. Large trucks can use liquefied natural gas (LNG), which isn't likely to be getting into your car's fuel tank unless you live somewhere very cold. This is a cryogenic liquid, formed when natural gas is chilled to around -160°C depending upon composition. Compressed natural gas (CNG) is more practical for cars. It can be stored on a vehicle in high-pressure tanks and CNG cars have typically between 60 and 80 per cent of the CO_2 emissions of petrol cars plus they emit much less NO_x. But they need bulkier fuel tanks and they take slightly longer to fill. Gas cars could be part of the short-term bridge to an all-electric future but in most countries they're unlikely to be a significant long-term solution.

The nuclear options

For every sane solution to the problem of emissions there are several outlandish ones too. Let's talk nuclear. Each kilogramme of enriched uranium is equivalent in energy terms to 2,700 tons of coal. So, nuclear cars could render altogether obsolete the whole idea of refuelling. A reactor of the right size to power your mum's Audi would need more enriched uranium only every five or ten years and there would also be no exhaust fumes. On the other hand, it would be difficult to switch off – perhaps requiring some way of siphoning off the energy to the national grid. And there's also the slight problem that the radiation emitted would likely kill the driver, passengers and any nearby bystanders. An aircraft carrier or large submarine is big enough to carry the required radiation shielding. With current technology, a car is not.

There were a few brave (or foolhardy) attempts at nuclear-powered concept cars in the 1950s and 1960s, though working prototypes were never built. The 1957 Ford Nucleon, for example, featured an interchangeable reactor that owners would swap every 5,000 miles and included the option to choose from different performance levels. France in 1958 saw both the Arbel Symétric and Simca Fulgur. Ford tried again in 1962 with the spectacular six-wheeled Seattle-ite XXI, but the interactive computer navigation and modular design were more accurate indicators of the future than its interchangeable fission powerplant. Perhaps the most optimistic concept from the atomic age is the Studebaker-Packard Astral. As well as being nuclear powered, it would balance on one wheel thanks to gyroscope technology and feature a force field to protect it against collisions.

The notion of the nuclear automobile occasionally resurfaces. Cadillac did a World Thorium Fuel or WTF (ho-ho) concept in 2009 featuring 100-year refuelling intervals of the thorium core; one gram yields the energy equivalent to 7,500 gallons of gasoline when excited by lasers. The fuel may have

William Clay Ford Senior, grandson of motor pioneer Henry, with Ford's 1957 vision for the atomic age, the Nucleon.

some future in power generation, but for the usual safety reasons it looks unlikely to be included in cars for the foreseeable future. Cadillac really designed it as a tongue-in-cheek fantasy, but the razor-sharp concept still looks incredible today and features six mini-wheels within each tyre.

At the moment, the only future for nuclear-powered cars is in fiction and games. The 1957 Nucleon is the inspiration for atomic cars in the *Fallout* video-game franchise, which has a variety of nuclear cars in its digital fleet. When shot at they explode in a mushroom cloud.

If nuclear is too far-fetched as a zero-emission option, what about air? It sounds reassuringly wholesome and is not actually as ridiculous as you might suppose. Air-powered trams were successful in parts of France in the nineteenth century, operating reliably for decades. Back in 2009 on *The Gadget Show* we filmed a car that ran on compressed air. It did only 30 mph but the claimed range was a passable 65 miles and there were no emissions to accompany the amusing noise it made, which was rather like a sort of hydraulic sewing machine. The styling wasn't exactly inspiring: the car resembled something between a slug and a snail. It had portholes instead of side windows, and it carried only one person – the driver – in what looked like a rather vulnerable position at the front. Nevertheless, its inventor, Guy Nègre, and his son, Cyril, had high hopes for the Airpod – as they'd christened their device, years before Apple used the same term for wireless earphones. Unfortunately, however, like most forays in the field, it's been dogged by technical difficulties and false starts.

Nègre's car stores its compressed air in a tank at high pressure and it powers a piston engine. However, the pistons are driven not by an ignited fuel-air mixture but by the expansion of compressed air, in a similar manner to the expansion of steam in a steam engine. Refuelling could potentially take place in a few minutes; there are no emissions and there's no need for a cooling system, spark plugs, starter motor or an exhaust. The car can even recapture some energy while braking, like electric cars, by compressing and storing air.

There are a few technical shortcomings, though. You need energy to compress air in the first place. And you also need energy to cool the air while compression takes place. This, together with low energy density, reduces the overall efficiency to around 30 per cent. Expressed in terms of energy per litre of space, the energy density is about one tenth of a typical lithium-ion battery, which is in turn, as mentioned above, less than a tenth of gasoline. Also, when it decompresses, the air gets very cold and it can literally freeze up the engine. In a dystopian world of global-warming apocalypse, it could at least form a ready-made air-conditioning system.

Guy Nègre licensed the technology in 2007 to the Indian giant Tata, which also owns Jaguar Land Rover. Photos have appeared of a multi-passenger taxi model. So far neither model has made it into production and the same seems to be the case with the restyled AIRPod 2.0, featured on a website operated by a US company called Zero Pollution Motors. Guy died in June 2016 aged seventy-five but his son continues his work.

Powering the future

Longer term, I think it is fair to say that we will increasingly move to batteries and hydrogen but that petrol and diesel cars will continue to justify their existence. We've been used to personal transport based on oil but the future will be based on diversity. We're going to have a greater variety of propellants: very clean internal-combustion engines, various degrees of hybridisation, battery-electric vehicles and fuel cell electric vehicles. These technologies will compete with each other in terms of efficiency and pollution reduction. With luck we'll have good unbiased information and let the market work rather than have governments dictating that 'it's got to be batteries' or imagining that simply throwing prohibitions around will be the only stimulus required. Instead, here's hoping that governments will adopt a more positive approach of intelligent regulation and help to drive synergies between industries, as in the case of hydrogen power. A blinkered 'give it up' message likely won't work.

Whatever the headlines, road vehicles are responsible for a smaller and smaller proportion of air-quality problems. According to Nick Molden: 'If we were to wind forward ten years from now, and if no other policies were put in place for cars, you'd find that cars were a pretty small proportion of urban particulates and NOx. They have had a dodgy record over the last fifteen years of failed regulation but they've finally fixed it for the very latest versions.'

Cleaner combustion engines and an increasing proportion

of zero-emission vehicles will start to challenge the automatic assumption that carbon emissions are caused by cars too. Authorities and industries will have to shift focus to address the other factors lying behind air-quality and carbon-emission problems.

THREE

DESIGNING THE FUTURE

How the Shape of Cars will Change

The car is the most significant and influential expression of industrial culture the world has ever known – even more so than the now ubiquitous iPhone. Back in the 1980s, the design guru Stephen Bayley – perhaps best known for resigning in a huff as creative director of London's Millennium Dome after a disagreement with Peter Mandelson – suggested that were Michelangelo alive today he wouldn't be wasting his time creating marble sculptures for tombs in the Vatican, but would be in Detroit working in brown modelling clay to create the modern world's true modern art form: the motor car.

Top of the tree in terms of manufacturing sophistication for more than a century, cars are the most complex mass-produced industrial products in the world. Each one has around 10,000 components, yet normally works perfectly for years on end – often while being mercilessly thrashed in the hardest and most arduous of environments. At their best, the standards of design that cars embody are impeccable, with interiors as good as product design anywhere and exteriors that can attain the sculptural qualities of fine art. Car technology and engineering are not only incredibly complex; they're also surprisingly affordable – cars are assembled to precise quality standards, hundreds of thousands of times over, to a price that, in the developed world at least, most people can easily afford.

Cars have, over the years, become vastly more varied, with new genres and niches appearing all the time. Sports cars first arrived during the Edwardian era, with the 'Prince Henry' Vauxhall, so-called because it was specifically designed to compete in a 1,200-mile-long 1910 motor trial named in honour of Prince Henry of Prussia, Queen Victoria's grandson. Civilian off-roaders bounded into view after the Second World War in the shape of the Land Rover and Jeep. Grand tourers, hot hatches, supercars, estate cars, people carriers, sports utility vehicles and crossovers between all of them have added to the daunting array of aesthetic choices open to today's car buyers.

On the way, some true design icons have been created. The 1959 Cadillac and the 1961 Jaguar E-Type are internationally recognised emblems of industrial creativity. Some of the most charismatic supercars ever designed soon followed – cars like Tom Tjaarda's Ferrari 365 California, Marcello Gandini's Lamborghini Miura or his iconic Maserati Khamsin. And star cars still emerge from creative designers today. The Alfa Romeo 4C, the Jaguar F-Type and the Ferrari 458 are just a few exemplary contemporary designs that look set to take their places on the automotive wall of everlasting fame.

Car-making culture has spread round much of the world, from Europe and America through Japan and South Korea to India and China. And with it there has been the rise of the car brand. Bristling with styling cues and a heritage that makes each one distinct, car marques are some of the most carefully honed, promoted and valuable properties in capitalism. Each has a meticulously defined style and appeals to a defined tribe of consumers, wannabe consumers and pure enthusiasts.

Conceptual art

Most full-scale future-gazing is done by manufacturers themselves with their concept cars. They perform various roles. One is pure fantasy. The heyday of American and European concept cars was probably the period from 1950 to 1975. Inspired by the Space Age, creations like the 1961 Ford Gyron – a dart-shaped two-wheeler that used gyroscopes to stay upright – weren't serious projects but they looked fantastic. The jet-inspired Oldsmobile Golden Rocket from 1956 and the GM Firebird III from 1959 are even more outrageous, both products of the aero-obsessed fertile brain of General Motors' styling chief Harley Earl. His rival at Chrysler, Virgil Exner, produced equally stunning visions, such as the 1956 Diablo or 1960 Plymouth XNR.

There's also a whole genre of spectacular fantasy supercars that were always too thin, pointy, impractical or fragile to quite make it into production. Consider the almost flat

Ford's 1961 vision of the future, the Gyron, was a balancing act: a two-wheeled dart kept upright by gyroscopic stabilisers.

1970 Ferrari 512S Modulo or the frangibly greenhouse-like 1967 Lamborghini Marzal. The definitive example of the breed is probably the still-stunning 1970 Gandini-styled Lancia Stratos Zero, which appeared in Michael Jackson's 1988 film *Moonwalker*.

The purpose of other concepts is to test out public reaction before cars hit the market, or to allow manufacturers to assess the reception given to styling themes before gambling the company's future on them. The BMW i concepts helped to prepare the public for the expensive move the company made in the direction of electrification. Another BMW, the Gina, which was created by the legendary designer Chris Bangle in 2001, used cloth bodywork that could be morphed into various shapes, transforming the car. (If you need a spoiler, for example, you can grow one.) Bangle claimed that his experience working on the Gina enabled him to massively and confidently extend the range of shapes that steel could be pressed into.

Some concept cars are surprisingly close to future production cars and generate interest in them. For example, the 1995 Audi TT prototype almost precisely predicted the real car apart from the rear windows; the Lamborghini 350 GTV mirrored the 350 GT; and the Jaguar CX-16 and CX-17 concepts closely foreshadowed the F-Type and F-Pace respectively.

A sprinkling of futuristic design surfaces in films and TV. *Minority Report* from 2002, for instance – that most polished of prophetic movies – includes a rather neat fuel-cell-powered self-driving 'Lexus 2054' with advanced biometric security features like eye and fingerprint recognition. The film's fast-moving living rooms on wheels that race down highways and climb walls look surprisingly like current autonomous concepts such as the Renault EZ-GO, albeit somewhat more sprightly. Not all futuristic movie cars are so inviting, though. They may be distinctive with projecting, spidery front wheels and vertical-take-off capabilities but I doubt you'd fancy a future where most cars are as gawky as the 'Spinner' cop cars in the original *Blade Runner*.

Film product placement is seen by some manufacturers as an opportunity to align their brand with a positive-looking future. For the 2019 animated Will Smith movie *Spies in Disguise*, Audi struck a deal with its distributors, Twentieth Century Fox, to feature the RSQ e-tron concept car used by spy character Lance Sterling. Naturally, it is fully autonomous when required and

Futuristic cars from the movies

A Clockwork Orange (1971)	The Durango 95 was a real car that looked outlandish enough for Stanley Kubrick's dystopian vision of the future. A mid-engined car created to explore the 'extremes of styling' by Dennis and Peter Adams in 1969, it's really called the Adams Probe 16. The car still exists and is so low it's almost impossible to drive. You can get in through the roof.
Minority Report (2002)	John Anderton – played by Tom Cruise – has the Lexus 2054 built round him in an automated factory. It's self-driving when desired, with hydrogen power, biometric security and voice control. Electronic rear-view cameras instead of mirrors were a novelty back in 2002 but are now relatively common.
I, Robot (2004)	Audi developed the gorgeous RSQ especially for the film. Top features include automation and wheels that are capable of moving in any direction. Many of the styling cues appeared on the next generation Audi TT.
Tron: Legacy (2010)	Created by German car designer turned 'automotive futurist' Daniel Simon, the Light Runner looks like a cross between a Formula One car and an off-roader. It could morph from a grid-runner into an all-terrain vehicle on demand. It's unlikely that the light-missile launcher would make it into a production motor.
Blade Runner 2049 (2017)	The flying police cars of the future are an update on the cars Syd Mead designed for the original film, and feature vertical takeoff, autonomous driving and detachable, gesture-controlled surveillance drones. The Peugeot badge might seem an odd choice for an American-made movie, but the company is eyeing a return to the US market and the movie's producers had high hopes for the film in China, where Peugeot-badged cars sell well.

features a holographic speedometer. It follows on from the company's successful involvement with *I, Robot* and *Iron Man*.

Will the future kill off our design heritage?

All these prototypes have helped to create a magnificent legacy of design culture around the car. But is this under threat? Some of the developments revolutionising the car industry today can be seen as good news. Electric cars, with their more flexibly shaped and compact powertrains, could free designers from a whole host of constraints and in turn stimulate more distinctive styling. But new trends also suggest trouble ahead. Car sharing and the rise of mobility as a service, rather than as a pure physical product, may require car makers to think more like airlines than like aeroplane manufacturers. Efficiency and comfort rather than looks may determine profit margins. And the autonomous future doesn't seem a great omen for good design if current driverless car prototypes are anything to go by. They're a rather depressing prospect, all too frequently blob-like pods with less allure than an airport shuttle. Just look at the slipper-like Waymo Firefly or the GATEway vehicles pottering around Greenwich in London. The average washing machine boasts a more assertive and exciting countenance.

For some this won't matter. When you're renting a shared autonomous car, its styling will be as irrelevant as the look of an Airbus or Boeing plane is to most flyers. What is crucial is the service you get from the airline or, in this case, the outfit that maintains the car and provides the much-needed mobility. The pride of ownership and self-identification with those carefully crafted car marques will be history. BMWs, for example, were once the Ultimate Driving Machines but if you're no longer driving or owning one you'll need something else to secure the brand's future.

In search of that alternative, and a future where we are all spared the fate of a miserably bland roadscape, I visited two of Britain's leading educational institutions, Coventry University and the Royal College of Art in London's Kensington. Between them they're responsible for training a massive proportion of the world's top car designers.

I wanted to see what these up-and-coming designers, and some of their tutors, had in mind for our motoring futures. If they were thinking positively then I could relax, confident that car design would continue to amaze and delight. If they had given in to the defeatism of pod-like anonymity then I would have to go into mourning as we'd be doomed to a future of lowest common denominators.

Coventry and the RCA don't have quite the monopoly on car design they once did. When I filmed an item on how to be a car designer for *Top Gear* back in the 1990s they were practically the only game in town. Now there are thriving courses elsewhere too, such as at the Strate School of Design in Paris and at the Umeå Institute of Design in Sweden. But taking an undergraduate transport course at Coventry and progressing – if you're lucky and skilled enough – to the postgraduate course at the RCA is still a route of choice into the profession.

The RCA's Vehicle Design course was renamed Intelligent Mobility in 2019, perhaps in recognition of the car's changing morphology. There are only twenty or so places available on it each year and it boasts a long list of influential alumni, including the current design chiefs of more than a dozen of the world's car companies. For starters there's Julian Thomson, design director of Jaguar (who replaced Ian Callum, another distinguished RCA graduate); Marek Reichman, design director of Aston Martin; and Peter Horbury, the man who reinvented Volvo in the 1990s and who's now design boss of car giant Geely in China. The heads of four out of five of Geely's design studios are graduates of Coventry.

Considering they're temples dedicated to the discipline of design, both courses are housed in surprisingly unattractive buildings. At Coventry they're in a concrete tower block next to the Ring Road; the RCA is a slightly more prepossessing concrete building in the shadow of the Royal Albert Hall.

To me the most difficult challenge is how anyone can get excited about designing cars we don't even drive. But it's not one that seems to worry today's students on either course; in fact I was struck by their zealous anticipation of an autonomous future and readiness to embrace new opportunities.

Driverless cars can be more 'architectural' – so enthused graduate Sam Philpott, whose Haven concept was inspired by the feeling of calm he experiences while relaxing in a window seat at home. The Haven is a vision of an urban vehicle for young people looking to reduce their stress levels; this would be privately owned but shareable by invitation.

A mindful window on the restless city and a study in mobile architecture, Sam Philpott with his Haven.

It looks rather attractive in spite of a distinctly rectilinear silhouette. A slope in the lower section is sufficient to express movement, Philpott thinks, and I am inclined to agree. It's a kind of blend between vehicle and building design. It also exemplifies the way autonomy will shift the design emphasis on to cars' interiors. Going driverless frees up the inside space in so many ways. At its most obvious, there's no need for a steering wheel and no requirement for a view of the road ahead. There's also no longer that sense of a hierarchy in the car, with the driver 'in control'. Philpott sees the Haven as having two different types of space: one for 'mindfulness' and another for 'interaction' with people invited in to share the environment. It even comes with its own sort of dating app, where passengers can be selected to interact with on stress-reducing urban meanderings.

This wasn't the only concept that explored the role of the car as an escape from the pressures of the modern world. Others I encountered included one that featured a meditation room made out of wood; one that featured plants and the sound of trickling water; and even one – perhaps for the more paranoid traveller – that was shielded against all radio communication.

Some designers even see the car as a fully fledged mobile room that when you're not making journeys in it becomes part of your house. The most ambitious car I saw, if indeed it can be called a car at all, was Sebastian Rokeberg's Polestar PS10. Don't let the Volvo sub-brand fool you – this is as much a house as a car.

Rokeberg's vision is a compact two-storey, luxuriously equipped home. It's about the size of two London double-deckers parked next to each other and can be part of a high rise or located independently in a remote landscape. Part of the living-room furniture is also a car interior – the 'interior module'. This includes three seats in different states of recline, a fridge with two compartments at different temperatures, footrests and tables. When you want to travel, you stay in your seat while the floor of the living room slides away and the module descends to meet a fully charged electric platform that contains the wheels, powertrain and batteries.

A section of the floor slides back above you to become the car's roof, which is supported magnetically to give an entirely open experience; if you don't fancy that, or the weather's not playing ball, interactive display screens roll up to shield you from the outside environment. The roof also contains the autonomous driving equipment. When you return home the process is reversed and the module resumes its role as part of your living room.

The interior module can be separated from the platform at certain hubs, too – at, say, restaurants or shopping areas; places where you can interact with others. It can also move round the house to take advantage of sunlight; to this end a maglev staircase can move out of the way as required. This may sound like something from *Wall-E* but the design is very tastefully done, honest. If it ever works!

Several of the nascent designers thought we could instigate a relationship with our driverless cars by emphasising all the things we like about cars that aren't the actual business of driving – such as a sense of your own space, or the physical sensation of movement and control. This might involve physically moving your seat or cockpit about to exaggerate sensations of cornering and acceleration, much like a fairground car-racing simulator. Or perhaps introducing virtual reality via headsets or screens, so that you can be in reality stuck in traffic at 30 mph but feeling like you're on an open country road doing 130. Air can even be harnessed to create passion by channelling it directly into the cabin to create a feeling of speed when you're not travelling that fast.

In place of the charismatic noise of the internal combustion engine, Irene Chiu advocates personalised soundscapes for autonomous-vehicle interiors: a car that creates its own audio experience. She'd made a highly engaging video of what a future trip in a self-driving Bentley might sound like. After a quick calibration to the driver's ears the car starts off in town, where some of the jarring noises are filtered out. As it powers through a winter landscape that looks distinctly Nordic, the natural sounds of that environment are magnified – including that of a sika deer crossing the car's path (of course, being perfectly automated, the car deftly avoids the obstacle). As night descends and the northern lights appear, the car creates eerily celestial music to enhance your journey.

You could choose to hear more or less of the ambient sound according to your mood, and you'd still be able to listen to music or the radio if desired. I presume the car would need a noise-cancelling system like that used in headphones, or like the in-car version that those with long memories might remember was pioneered by Lotus back in the 1990s. Chiu envisages conductive sound through the seats as a way of enhancing the travel experience for deaf passengers.

Another suggestion to enliven the autonomy is to combine it with some sort of vestigial driving experience in the same vehicle. But I'm not convinced. The notion of a dedicated driver zone in an otherwise autonomous car seems too wasteful of space and resources. I think you're more likely to have a different car that you rent for a track-day experience than you are to spend the extra cash on a car that can do both.

And in any case such designs appear to make the car far too large for ever more crowded road conditions.

None of the designs seemed to have the fluidity of film cars in this respect. In *I, Robot* Will Smith's Audi R8-inspired hybrid switches quickly and elegantly from self-driving to driver mode with a smooth extending steering column. I think it's one of those transformations that will always look better in the movies with a wealth of CGI at their disposal.

To share or not to share?

Autonomy brings with it the question of whether future vehicles will be shared or owned, or a mix of the two. For many drivers, sharing will be a horrific prospect. One of the traditional joys of the car is that you're in your own private space in a crowded world, with your own media and even to some extent your own climate, a locale shared only with those to whom you've given express permission. A shared car has more in common with public transport: dirt, inconvenience and the great unwashed threaten at every turn.

Whether you'll accept sharing will depend upon where you live and how old you are. In America autonomous vehicles look far more likely to be private and personally owned. In Paris the authorities have already decreed that all autonomous vehicles must be shareable by law: self-driving cars will be required to scurry away by themselves to be used by someone else.

According to one Coventry student, Denny Julian Deemin, Generation Zs (born between the mid-1990s and mid-2000s) will still want their own cars but Generation Alphas (born since 2005) will want a simpler lifestyle without the bother of ownership. When buying cars, particularly used ones, I've always found the hassles of dealing with insurance, registration documents, finance and provenance checks rather onerous, so I can to an extent empathise with Denny's generation Alphas. However, I wouldn't yet be ready to go as far as abandoning car ownership altogether.

Idealists as they are inclined to be, the students saw lots of advantages in sharing. Some notions were more like small-scale buses, trains or shops on wheels. Concepts included a massage parlour, a sort of mobile-games console pod for entertainment, a doctor's surgery and the inevitable pub. To ease the potential pain, technology will help you decide who to share with – like having the equivalent of Tinder to suggest compatible travelling companions. As most of us like listening to music in our cars, Or Shacar from the RCA proposes sorting your fellow occupants by Spotify playlist so at least you'll be listening to the same tunes.

Another idea is cars that are a mix of shared and private spaces. These could be useful for people who don't want to share but who need to when they travel into city centres where single-occupancy vehicles have been banned. Just like swarms in nature, these autonomous vehicles could be continually changing in size and configuration. RCA graduate Yang Liu envisages a car with several distinct sections that can operate either independently or together. A front section is for group sharing; and I rather liked the way Liu's seats were composed of separate inflatable sections so they could adapt to different sizes of passenger or deflate completely to carry someone in a wheelchair. The back of the car has an individual area with greater privacy, while at the very back are three smaller autonomous pods – two of these can either travel with the car or break away for local deliveries, while a third provides charging and cleaning on the move so the whole system never needs to stop except to load and unload.

It's an exhausting concept just to think about, a world of endlessly varying perpetual mobility and autonomy within autonomy, but Liu is adamant it's efficient, making the best use of road space, time and energy.

Other shared-use concepts are more like hire cars. You still have your own space but you're renting it. Fukhita Wong from Coventry has created the Snooze, an autonomous mobile hotel that promises to combine the comfort of a conventional hotel room with the convenience of a taxi, bookable through an app that would hopefully offer guidance on where to park to avoid the noise of passing traffic. You can also use it to travel as you sleep in an individual incarnation of the train sleeper car.

A benefit of shared autonomy that for some may help in the pill swallowing is that there'll be no need to choose a car that can do everything – one that's equally at home on the urban commute as on holiday. There can be specialised cars developed for every conceivable role, potentially leading to more interesting designs. You can have different cars for different purposes delivered to your door.

RCA student Anand Askinar's Fractal is a shared vehicle specifically designed for families with a single young child or baby. It's a mobile space that will appear at your home on demand and can be configured as required. You can switch between a 'care mode', with a crib between the seats, and 'share mode', where you're chatting and interacting with an older child. The concept includes a buggy, which is separately autonomous if you want it to be, and a digital screen in the roof offers entertainment. As your family expands, or when you have visitors, you will presumably be able to choose from more commodious Fractals. However, having your precious offspring speeding along the motorway in an autonomous buggy in the company of trucks, buses and other vehicles would of course take parental anxiety to a whole new level.

Maybe you'll just carry round your own intelligent steering wheel. That's the idea behind Jaguar's Future-Type concept, first shown in 2017. You summon up different Future-Types depending on your transport needs, slotting the wheel into each vehicle. When you're taking the kids to school, for example, you might choose one with '2+1 Social Seating' so you can chat face-to-face while the car autonomously drives you there, using its 'Sayer' AI (named after the designer of the E-Type, Malcolm Sayer). Keeping all options open, Jaguar also envisages you perhaps owning a Future-Type as well and using your steering wheel for actual driving.

For me, some of the most exciting and romantic shared-vehicle concepts are the ones aiming to take the place of your holiday hire car by forming their own mobile tourism experience. RCA student Stavros Mavrakis has created the Zephyr. It has no relation to classic Fords with the same name, though it does have the same mythological and poetic-sounding inspiration: the Greek god of the west wind.

The Zephyr is designed to help you experience the Cyclades islands without the hassle of driving yourself round the narrow roads. There are design themes that echo the geometry and colours of local architecture; and local materials like rattan, a palm that can be woven into fabric, cover the cushions while the seats are finished with cork. The body shape is inspired by local boats and, in a chillingly modern twist, the hull is made from a composite material derived from ocean waste.

'You're not just being transported; you're absorbing the experience of the region by using this car. It's the key part of an experiential adventure,' enthuses Mavrakis. With the help of his VR headset I was able to imagine the relaxing autonomous experience of being transported around an idyllic island. The view of the local sand and boulders would be framed by my land yacht's hull, while I rested my feet on a rattan cushion and immersed myself in ancient mythology. As all this goes on, the built-in turbine would be harnessing the very wind after which my craft was named to charge the batteries ready for my odyssey to continue. Perhaps the remaining seven of the eight Greek wind gods could inspire different, equally characterful tourist conveyances.

Luca Robert Barberis's Infiniti concept is an autonomous-car subscription service specifically designed for tourism. The only thing you own is a custom-made suitcase that fits into a slot in the bonnet. The car is low and muscular; it's like the best classic sports cars from the front, while from the rear it has the elegance of a traditional coupé. From the top there's a teardrop profile with crease lines and roof graphics to emphasise the dynamism. It has real glass rather than screens, so you can see the actual view, but augmented reality tells you more about what you're seeing. Barberis, who's from the US, thinks too many of his country-folk are geographically ignorant and wants to change this. 'You come back from your vacation a better-informed person.'

This service would be at the premium end of the vacation market. There'll still be a role for carefully nurtured upmarket brands in an autonomous world but it'll be geared to a high-end service. According to Professor Dale Harrow, who's head of the Intelligent Mobility Design Programme at the Royal College of Art in London, 'It's not just about materiality; it becomes the whole experience of the journey from the moment you leave your home to when you arrive somewhere, including how the vehicle presents itself to you at one end and how you say goodbye to it.'

Harry Wilson, a Coventry student, told me about his project called the Usu, where you subscribe to an upmarket brand that would become a sort of universal global taxi service giving you access to a fleet of identical autonomous cars. Whether you're in London, Shanghai, New York or Singapore you'd interact with the (inevitable) app on your phone and configure the car as you wanted it before you arrived. When you reach your departure point the car's already there with seats and lighting set to your preferences and your choice of entertainment playing. If you're an international business traveller and you left this car at Heathrow, you'd get an identical car coming round to collect you in Shanghai and drive you around the city.

As someone who has long been annoyed by the appalling service offered by airport hire-car companies, with their penchant for giving you a car of the wrong size, renting the one you booked to someone else and charging mind-blowing excess payments, this would be an understandably appealing service in the autonomous era. Autonomy of course is a long way off. Before we get there, the rise of electric cars is the most important development.

Should electric cars look different?

An immediate issue is what electric cars should look like. David Browne, who was director of the Coventry course for two decades, questioned whether there will be a distinct design language for electric vehicles, noting that cars like the Honda Insight, with its teardrop shape and enclosed rear-wheel arches, the original Nissan Leaf, and the BMW i3 and i8 have made a stab at creating a new look for electric cars. 'What I liked about the original Honda Insight, and now like about the i3 and i8, is that they didn't take their conventional car and put a bunch of batteries in. They're engineered to be what they are. The Insight was aerodynamically efficient and used exotic materials to keep the weight down. The i3 and i8 were designed to be electric and not spin-offs from any other BMW. The production and the mechanics are different, as is the styling.'

At the RCA, student Bin Sun from China sees the Insight and the BMWs as exceptions to the rule and set himself the task of creating a new design language for electric vehicles. 'At the moment the electric car mainly looks the same as an ordinary car. Aggressive. A fake grille. Angry eyes. But I believe we have some unconscious feeling about electric power. Petrol power is furious and sometimes even dangerous. Electric energy is purer and safer. It's more rational and peaceful. When someone drives an electric vehicle they say how quiet it is. My main attention is to express those feelings.'

I suggest that people also often say how quickly an electric car accelerates, so it's not all about a new humility, but have to concede he has a point. When I drove a 2018 Nissan Leaf for a short period I couldn't help noticing that many people expressed their disappointment to me that an electric car looked very much like any other car.

Bin describes his concept as looking more digital, with a central bar round the car and small grilles where the parts of the car that are being cooled, like the batteries, are located. The surfaces and graphics are designed to communicate what he calls a more rational line, with larger radius curves to promote a feeling of calmness. A spectacular pair of gullwing doors adds some glamour.

Dongwon Lee came up with a suggestion for Tesla, a company admired for its technical expertise by the students but less for its design, which was considered rather bland. His Project Berry is mixture of the nostalgic and the fantastical. It's part car, part spacecraft, a sort of concept car for space exploration.

Addicted to space himself, Dongwon thinks Tesla should let some of Elon Musk's SpaceX project genes loose in their cars. His one-seater was inspired by 1960s concept cars. 'Space was fresh and new. It was the age of spaceships, with thoughts of people moving to space, and space movies like *2001*. Design in the sixties had space-orientated shapes, whether in cars like the GM Firebird or furniture like Eero Aarnio's ball chair. They're soft. Each part is interconnected and seamless. My rear lights and front lights are linked in one circle, for example.'

Fancy a trip to Mars? Electric cars like Dongwon Lee's 'Project Berry' could revive the space vibe.

The car has the look of a personal USS *Enterprise* about it. I wondered about the large overhang at the back but that's apparently space-inspired too, from the tails of comets and the shape of Saturn's rings. It's a bit like a Harley Earl or Virgil Exner concept for the twenty-first century.

Honda looks set to continue the electric differentiation with its Sports EV concept and Honda e production car, which have a sort of innocent and wide-eyed, circular-headlamped and round-bumpered look. But other manufacturers are committed to a more conservative approach. Tesla says it'll keep its styling conventional to maximise market appeal. BMW has expressed a desire to make its new electric cars more like its conventional models, while VW's electric I.D. concepts look very conventional apart from their full-width LED lights and sliding doors (which apparently won't make it into production anyway on cost grounds). In the short term I think conservatism will win out and we'll see electric cars merging with the mainstream. But, in the longer term, brands are always going to be looking for new ways to stand out.

The future of status

Status and the car have always been closely linked. Not for nothing does the famous car stylist Chris Bangle, who was responsible for reinventing the look of BMW in the 1990s with his flame-edge effect, call the car an avatar. It's a reflection of yourself: you buy it and you become it.

For the first few years that cars existed, the status attached merely to owning one meant that it didn't really matter what they looked like. But it didn't take long for prestige cars to appear. The human desire to look different and be seen above one's peers in the pecking order quickly extended to motoring. A 1904 Mercedes was a vastly more prestigious device than a De Dion-Bouton of the same year. In 1906 the new Rolls-Royce Silver Ghost was justifiably marketed as 'The Best Car in the World' and created an ultimate status symbol in the process.

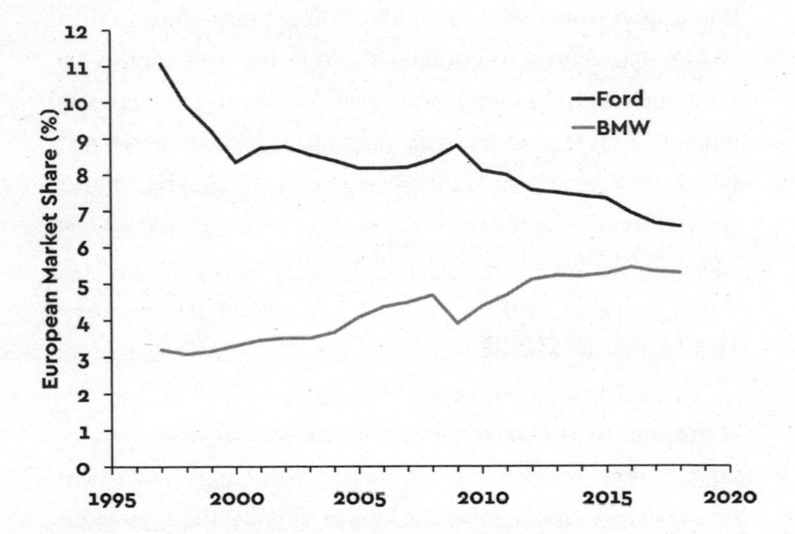

Keeping up appearances: over recent years luxury brands have boomed.

By the 1920s the likes of Delage and Delahaye in France, and Duesenberg, Cadillac and Pierce-Arrow in the USA, were producing superbly crafted and glamorous conveyances out of the very highest-quality materials. And if that wasn't enough, you could buy the chassis and commission a coachbuilder to create something even more extravagantly individual.

Style spread downmarket in the 1930s, with streamlining applied to more affordable cars from the Lincoln Zephyr to the Flying Standard. In the 1960s, marketing departments refined the buying process by multiplying the options available to ordinary car buyers. The 1964 Ford Mustang was available in a groundbreaking number of variations. Ford brought the trend to Britain with the 1969 Capri, which featured twenty-one different engine and trim combinations and was sold as 'The Car You Always Promised Yourself'.

Everybody wants status. Helped by leasing deals, sales of upmarket brands have boomed at the expense of the mid-market. These days the top-ten sales charts in countries like the UK are more likely to include BMWs and Mercedes than they are Ford Mondeos. More luxury cars are sold than ever. Look how sales of Ford-badged cars in Europe have declined over recent decades, while those of BMW and Porsche have soared. It's the same story in the US.

Status will become easier to attain thanks to new technologies like scanning and 3D printing, which will facilitate the bespoke manufacture of cars. They will foster a revival of small-scale production and even the individually commissioned car. Coventry 3D printing company Vital Auto already prints complete headlamp assemblies for the NIO EP9 supercar, which is assembled in China. The company sees huge potential for 3D parts for cars with small production runs of between one and a thousand per year.

This rebirth of bespoke will be helped by the move to electric platforms and the 'skateboard' chassis. This is a compact floorpan that incorporates all the batteries and powertrain and unsurprisingly looks rather reminiscent of a (very large) skateboard. It will be relatively easy to create custom designs on this simple, common chassis. Semi-autonomous crash-avoidance features like the automatic emergency braking already seen on current mass-produced cars will also help to keep the car safe without the need for expensive crash testing and the associated complex body structures.

Brian Clough is a former senior lecturer at Coventry and now a freelance designer who is keen on the potential for customisation. 'It's a throwback to the days of Mulliner Park Ward and other coachbuilders, where people put bodywork on a chassis. We might see a democratisation of bespoke. You and I could sit down at a tablet and agree a design on Alias modelling software and produce a 3D model within a week, in the way you hire an interior designer to decorate your house. You'll sit there with a designer and design a car to suit you, they'll make a little car model for you to evaluate and a month later you'll have the full-size one sitting on your drive. The only limit is your imagination.'

New kids on the block

Creating more upmarket, individual cars may become easier but the old brands aren't going to have it all their own way. A new force in the market is China. At 23 million units a year it is already the biggest car producer in the world. Its factories crank out more cars than the EU, or the USA and Japan combined. It has 70 different car manufacturers and a bewildering variety of its own brands, though many of the cars sold there still have Western badges on the bonnet. They are produced under joint arrangements with the Western companies that helped establish the industry in the 1980s, but that are not presently allowed to produce cars there in their own right, either for the domestic market or for export.

There have been some issues with Chinese models, ahem... 'taking inspiration' from other well-known car designs, most notably the BMW X5 (Shuanghuan SCEO), Porsche Macan (Zotye SR9) and Range Rover Evoque (Landwind X7). Perhaps more surprisingly, other targets have included the likes of the Daewoo Matiz (Chery QQ), the Smart Fortwo (Shuanghuan Noble) and even a Mercedes-Benz CLK front bolted on to a Renault Megane rear (the BYD F8 Coupe). Slightly less comprehensive imitations include the Mini knock-off Lifan 320 and the Renault Twizy-esque Rayttle E28. You can even buy kits to replace the badges on your Chinese copies to make them look even more like the original.

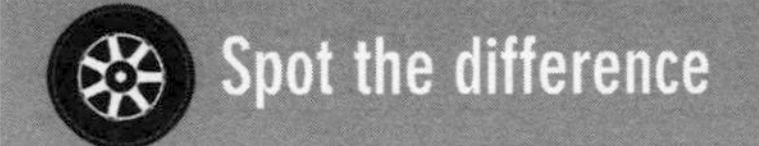

Spot the difference

Chinese imitation | **Original**

Zotye SR9

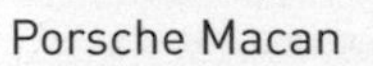

Porsche Macan

Landwind X7

Range Rover Evoque

Shuanghuan Noble

Smart Fortwo

Such misbehaviour will have to stop if the Chinese companies are to sell their cars globally, and do for cars what Huawei, Oppo and Xiaomi have achieved with smartphones. But, there are already signs that the Chinese courts are enforcing intellectual-property rights more aggressively. In addition, car manufacturers will need to improve quality – which has so far fallen well short of the standards reached by their counterparts in the telecoms industry.

Nevertheless, we'll be seeing and hearing a lot more from the big Chinese car manufacturers in the near future. Geely's prestige brand Lynk & Co is leading the way with exports and will be one of the first lines to make it into Europe and North America. Geely already exports to South America, Russia and the Middle East under its own nameplate and looks set to exploit its control of UK sports-car manufacturer Lotus by broadening the range of cars carrying the racing brand. The Geely Icon, first seen at Beijing Motor Show 2018 and based on the XC40 made by Volvo, may be produced in Europe.

New names in the West will also include Hongqi, best known for producing China's state luxury limousines but now promoting its GT Concept, a coupé somewhere between Bentley and Aston Martin in style. Dongfeng has shown some very handsome sports-car concepts, while XPeng concentrate on battery EVs. BYD is another big name that looks set to start exporting soon. The company has hired an ex-Audi designer Wolfgang Egger and unsurprisingly produced some rather Audi-like concepts. Qoros and Changan do more conventional-looking SUVs and aim to sell in the West in the near future;

and at the more prestigious end of the battery-electric market the NIO EP9 electric supercar scored one of the fastest times up the hill at the 2018 Goodwood Festival of Speed.

Africa will also become much more important. As things stand, it's the only major continent on earth without a substantial car industry. By 2030, however, it is expected to be the fastest-growing car market in the world and will create another location for manufacturing. The 'digitisation' of production will make the location of factories and their role as important national icons less important than the software making the car and controlling its production. According to Brian Clough, 'Now, you can almost open a factory anywhere. It's not as difficult as it was in the days of the Hillman Imp and the Vauxhall Viva to take production somewhere else. As in the aircraft industry, you don't have to make the part in a particular factory anymore; you just send over the data.'

It's a development that has parallels in the RCA and Coventry students' own model-making. Virtually all the projects are built as scale models just under a metre in length. Traditionally students have produced these themselves using clay, a skill that has been one of the definitive talents of a car designer and one that's in demand from other industries as well for the sculpting expertise it represents. While Coventry's students were still creating their own clay models, in London this was the exception rather than the rule, with designs taking place mainly on the computer and the file sent elsewhere (usually China) for 3D printing into a resin model. 'Made in China. Designed here, though,' noted David Browne.

The old meets the new

Clay is unlikely to die away, though its demise in this role was thought inevitable until recently. Some students said they emphatically preferred working in clay and couldn't truly work with shapes on screen. As Brian Clough confirmed, 'Renault was one of the companies trying to get rid of clay modelling entirely. And make it all digital.' But it changed its mind. 'There are still lots of companies that want to stick with clay models because the way you perceive a three-dimensional object is different from the way you see it on screen, even using the best VR systems. Clay modelling isn't dead yet.'

The best of both worlds is to have both clay and digital. There's no doubt that looking at a 6x6-metre Powerwall – a high-definition video display made seamlessly of other large displays – through 3D glasses gives you a very realistic view of a new design, and one you can refine, tweak and re-colour in a fraction of the time it takes to craft another clay model. But clay itself is delightfully tactile and there's nothing like the realism it engenders and the relationship it creates with a design, which can be developed and refined as it's shaped.

A reverence for the past is likely to be important in car styling too. In the 1990s and early 2000s, cars like the Beetle, the Fiat 500 and above all the Mini all nodded significantly to their ancestors. The trend will embrace more recent cars. Even companies that would never have considered retro styling have started to plunder their past. At the Paris Motor Show in

2018 Peugeot showed their E-Legend, a brilliantly handsome reimagining of the 1960s Peugeot 504 Coupé by the company's chief designer Gilles Vidal. It's a car that the firm's CEO, Jean-Phillipe Imparato, says he would love to put into production after 2020.

Cars from Japan in particular are receiving much more retro attention, as the old ones become more fashionable. Student Seok Hyun Kim, for example, wants to revive the 1967 Mazda Cosmo in his Cosmo Revival 2027 concept. 'The original car challenged convention, with its strikingly individual styling and its rotary engine,' he says. 'The successor will carry on the spiritual heritage.' He envisages it lending its model name to a new premium electric sub-brand for the company. As well as echoing the old car's styling his car will continue to be technologically innovative, as demonstrated by its hydrogen-fuelled rotary-electric powertrain.

Even the truly kitsch will get a look in. Paul Busuego's Cadillac Emperado concept is an enormous luxury car with styling inspired by 1970s Cadillac Fleetwood formal limousines, including their rich burgundy interiors. It has a hugely long wheelbase to maximise passenger space, a built-in bar, which works like a Sushi conveyor belt, and champagne flutes laid out as a sort of V8 facsimile under the bonnet at the front. 'It's an extravagant and ostentatious way of displaying your wealth, more old-school American luxury than how many gadgets you can stuff in your car.'

Rather than fretting about rights and licensing, car companies usually welcome students using their brands in design exercises. They're intrigued to discover what creative young people, unconstrained by the quotidian formalities of corporate cultures, think about their cars' future. They even offer to help the students develop their ideas. These days, though, it's usual for students to ask the companies for permission first to avoid any potential legal tussles.

Sometimes there are more formal collaborations between design courses and car companies' design departments. In 2018 Bentley, for example, challenged second-year post-graduate students at the RCA to develop concepts around 'the meaning of future British luxury in the year 2050'.

Bentley's design director Stefan Sielaff, himself an alumnus of the RCA course, explained the company's motives. 'We are looking for ideas and concepts that could potentially lead us in new and interesting directions, using the perspective of these digital natives – from all over the world – to see things differently... the taste makers in training, if you will.'

As well as an appreciation of the old, new materials will undoubtedly become more common in car design. Carbon fibre and lightweighting will be crucial. Safety and equipment demands have meant that cars haven't enjoyed the 'dematerialisation' of other industries' products. Beer cans have fallen in weight over the years. They were 73 g in 1950 and 21 g in 1972 and are only 13 g today, according to Kevin Kelly's book *The Inevitable*. But cars haven't. A 1922 Austin Seven weighs 320 kg. A 1974 VW Golf weighs 790 kg. A 2019 basic Golf is 1,245 kg. If cars are to maintain large ranges and run on electric batteries, they will have to get lighter in the future.

Car interiors will also transform as a result of the increased focus on autonomy. There is likely to be a shift away from wood and leather as the go-to choices to represent luxury. It's already causing some friction in the world of established car manufacturers. In particular there was a ding-dong between Lagonda and Bentley after the 2018 Geneva Motor Show.

Aston Martin, whose luxury-focused Lagonda brand had remained dormant for years, revealed its Vision Concept, a Rolls-Royce-sized, all-electric, aerodynamic limousine to be produced in its new Welsh facility from 2021. The dramatic body was styled by studio head Marek Reichman. There were promised self-driving features, a retractable steering wheel and 360-degree spinning seats. This was to be expected; although Aston Martins should always be driven, the Lagonda brand is more at home with the notion of autonomy.

But more controversial was the interior, imagined with the help of English furniture designer David Linley. He'd blended

Silk, ceramics, cashmere and hand-woven wool seats redefine luxury in the Lagonda Vision.

ultra-modern materials with the traditional. The modern included carbon fibre on the trim panels and ceramic tiles that open and close to alter the ventilation and adjust the volume of the music. The traditional element comprised hand-woven cashmere upholstery and silk carpets. There was no wood and leather to be seen.

Reichman, himself an RCA alumnus, has been quoted getting a dig in at his luxury competitors. 'Rolls-Royce and Bentley are Ancient Greece today. I worked on the original Phantom. The brief was Buckingham Palace on wheels. It was important to do that to establish it. But the world has changed and the royals have changed [too].'

The chief executive of Rolls-Royce Motor Cars, Torsten Müller-Ötvös, hit back saying that the sports-car maker had 'zero clue' about attracting rich motorists and that 'they really don't understand our segment, they really don't understand the customers'. He was 'sorry to be so blunt'.

Such squabbles aside, another aspect of car interiors that is likely to transform in the future – and this is one that I'm particularly partial to, as a presenter on *The Gadget Show* – is the technology inside.

Gadgets!

No new motor-show concept car is complete without the ever-increasing scope of touchscreens. Buyers are certainly swayed by 'infotainment'. In a study based on 2015 data,

McKinsey found that 37 per cent of car users would switch from their current car make to another if it was the only one to offer full access to apps, data and media. Meanwhile 32 per cent of respondents would be willing to pay for these features on a subscription basis. The passion varies between territories. In China 60 per cent would switch manufacturers to gain connectivity features whereas the figure in Germany was 20 per cent. The younger you are and the more urban your abode, the more likely you are to make the switch. It should be a big deal for car manufacturers. Most customers are aware of, but seemingly not worried by, data-privacy implications of people having their driving data. However, McKinsey found that people were more relaxed if the apps involved relate directly to driving.

In terms of who controls this area there's going to be an interesting battle between car manufacturers, tech companies and maybe some altogether new players. At the moment bringing your own phone into a car and using Apple CarPlay or Android Auto is far better than trying to wade through car manufacturers' built-in infotainment systems. These are often confusing, difficult and costly to update. Also, given that the media is likely already on your phone anyway, it can be far too slow to load.

However, the 'bring your own device' model may prove inadequate in a world of more sophisticated connected car features. I suspect manufacturers will group together and offer certain features in conjunction with technology companies, forming a few competing car-related ecosystems, much as Google's Android operating system supplies several different phone manufacturers. These will encompass new features like real-time parking information, self-parking facilities and even collision avoidance. Developing applications for such ecosystems may prove as attractive to car manufacturers as it is for existing mobile platforms, giving car users access to the work of hundreds of thousands of motivated software creators.

I already want a sort of multi-window approach in Android Auto or Apple CarPlay so I can have Spotify or a podcasts app open on one screen and Google Maps or Waze on another, while a further screen is devoted to the manufacturer's in-car offerings such as terrestrial radio or the heating system. Indeed, one could almost imagine every surface in a car's interior being a screen, some of them interactive, so that while you're watching it, it'll be watching you for any commands, or to monitor your health or attention levels.

You may not even need to touch the screens. Augmented-reality systems might work with your smart contact lenses so that you can swipe through your heating controls with a gesture rather than have to make smudgy dabs on a screen.

The RCA's Professor Dale Harrow thinks that the future may well see a backlash against all this in-car technology. 'There's a lot of debate going on about the amount of information presented to you in cars. One thought is that a deep digital detox is required. You maybe should keep the car as a clean space. We're seeing lots of screens appearing – but the opposite of that is to leave them out and make it a very calming space.'

It will still be (mostly) about styling

Whether they're manufacturers' concept cars or student designs, the most memorable models that I've seen are the ones with that most traditional quality: kerb appeal. Perhaps above all else, car design is still about that emotional draw of a wonderful shape – a matter of surfaces and shapes and how light falls on them. As Coventry's design-course director Aamer Mahmud put it, 'It's got to be something that's so beautiful that people can't resist having it.' For me, three of the students' designs particularly embodied such values.

Arash Shahbaz's engaging update on the concept of an MGB mixes the modern and traditional in unexpected ways.

Arash Shahbaz has transported the MGB into the twenty-first century with a blend of tradition and modernity.

The shape of the car is cutting-edge modern, with a residual wraparound windshield, yet it manages to evoke the volume, proportion and stance of the MG original and even hint strongly at its two-slot radiator grille. And, remarkably, it would use bold, dominant leather straps to fasten the bonnet and echo earlier MGs from the 1930s. There would be no self-driving kit here. Power would come from four separate electric motors with full independent torque vectoring, so it's designed with both the road and the track in mind.

Shyamal Kansara showed me his Eclipsx. With wheels at an extreme camber and highly aerodynamic bodywork based round a flipped-over aeroplane wing it would, he claims, be as frisky as a fighter jet yet generate between 6 g and 7 g of downforce for extreme cornering – significantly larger forces than a current F1 car at maximum speed. It's a mesmerising-looking shape, though the tyre wear has yet to be estimated and may prove an issue. So far, the scale-model toy version he's built shows its agility. It can certainly turn on the spot; and he thinks that as well as performance applications it could help the elderly get into more ordinary cars by flattening the wheel arches, and also inform more efficient bus designs.

Christopher Saetrang's reimagining of an Alfa Romeo, the Milano 2027, is technologically advanced yet packed with heritage, showing influences from open-wheeled Alfas of the 1930s with their 'narrow grilles and open wheels', while keeping elements of the delightfully pretty 1960s Spider Duetto with its boat-like rear body shape. Any shortage of power encountered when putting his chosen MV Agusta motorcycle engine in a car body has been overcome by supplementing the petrol power with electric motors and supercapacitors. Everything from the bumpers to the roof – and even larger parts of the bodywork – would be fully customisable to maximise customer choice and also generate extra income for the company. There's a Formula One-style legs-in-the-air driving position but a sliding seat mechanism ensures that the car is still easy to get in and out of, in order to avoid alienating older buyers.

These three cars, all beautiful and desirable but incorporating massively different levels of technology, make you confident that the best days of car design are anything but over, and there's a good chance we'll see a truly exciting and engaging visual feast on the road ahead.

FOUR

SPEEDING AHEAD

the Future of Performance Cars and Motorsport

In a quote that surely applies more to the car than to any other means of transport, Aldous Huxley once famously said: 'Speed, it seems to me, provides the one genuinely modern pleasure.' Certainly the car's capacity to offer that thrilling sensation, magically under the driver's control, has proved more seductive than any of its other features.

But for every Toad who loves nothing more than driving fast on the road, captivated by the blur of foliage and tarmac being sucked past the windows, there is a nannying puritan who regards speed as the ultimate form of selfish waste and a pointless danger to others. Here, prejudice meets genuine concern and it's not always easy to separate the two.

Such naysayers haven't stopped the fastest cars becoming ever faster. In 1949, the Jaguar XK120 held the production-car speed

Fastest production road cars since 1949

Year	Car	Top Recorded Speed
1949	Jaguar XK120	125 mph
1955	Mercedes-Benz 300SL	151 mph
1959	Aston Martin DB4 GT	152 mph
1963	Iso Grifo GL 365	161 mph
1965	AC Cobra MkIII 427	165 mph
1967	Lamborghini Miura P400	171 mph
1968	Ferrari 365 GTB/4 Daytona	174 mph
1969	Lamborghini Miura P400S	179 mph
1982	Lamborghini Countach LP500 S	182 mph
1983	Ruf BTR	190 mph
1986	Porsche 959	198 mph
1987	Ruf CTR	213 mph
1993	McLaren F1	221 mph
2004	Koenigsegg CCR	241 mph
2005	Bugatti Veyron EB 16.4	254 mph
2007	SSC Ultimate Aero TT	256 mph
2010	Bugatti Veyron 16.4 Super Sport	258 mph
2017	Koenigsegg Agera RS	278 mph

record at 124.6 mph. By 1968 the Ferrari 365 GTB/4 Daytona achieved 174 mph (I can recall being very excited by this as a seven-year-old). The McLaren F1 scored 221 mph in 1993 and held the record for eleven years. Manufacturers have seriously pushed limits in their efforts to outdo each other in flat-out speed, and these record-breaking cars represent some of the finest achievements the world of motoring has to offer.

The latest high-performance 'supercars' and 'hypercars' are the fastest yet. The new American outfit Hennessey Special Vehicles unveiled its upcoming Venom F5 in Las Vegas in 2017. With its bespoke 7.6-litre twin-turbocharged V8 engine it's aiming to achieve 301 mph and 0–186 mph in under 10 seconds and 0–249 mph in under 20 seconds. The production run is to be limited to 24 and buyers will, apparently, be 'hand-picked'.

Providing you can stump up the asking price there are more cars than ever capable of hitting 200 mph, including convertibles like the Lamborghini Huracan, Ferrari 488 and McLaren 570s Spider, and even four-seaters like the Ferrari GTC4 Lusso, the Aston Martin Rapide and the Bentley Flying Spur. For those with an eye for a bargain the £89,900 BMW Alpina B5 can also hit 205 mph.

The Koenigsegg Agera RS holds the title of world's fastest road car at the time of writing. The Swedish manufacturer's one-off £1.5 million car clocked up 277.87 mph on a two-way run on a closed public road in Las Vegas in 2018. The technology is conventional but powerful: the RS's 5.0 litre turbocharged V8 produces a fearsome 1,341 bhp.

The popularity of supercars and hypercars – or whatever you

call these elite expressions of sculptural beauty and blistering power – is surging (among certain very well-heeled drivers at least). Bolstered by rising income inequality, the growth of new wealthy classes round the world and their seemingly insatiable desire to own something fast, supercar sales have soared in recent years. But can these vehicles survive an age where environmental and safety concerns seem to get more pressing by the day, and where any sort of wanton performance and speed is frowned upon?

In search of some answers I drove down to see Simon Saunders of Ariel. The firm took its current name in 1999, after a pioneering bicycle and motorcycle manufacturing company created by James Starley in Coventry in 1870. It's most famous for its highly individual, extremely lightweight exoskeletal sports car, the Atom. It's one of the fastest-accelerating cars in the world and a special V8 in 2012 could reach a top speed of 168 mph. Ariel also produces an off-road buggy based on the same design principles as the Atom, called the Nomad, and a 'naked' exoskeletal motorcycle called the Ace.

Those of a cruelly critical disposition have likened driving Ariel cars to driving a piece of scaffolding, but I've always found them tremendously exciting as both driver and passenger. Ariel operates from a small factory that appears to be literally bursting at the seams, located in Crewkerne, Somerset, many miles away from Britain's traditional motorsport corridor. The place looks like a large brick farmhouse built round a court-yard. I inspected the bright, well-disciplined manufacturing room, where one individual is responsible for building each car and bike from start to finish and puts their name on the completed vehicle. The atmosphere of quiet dedication there was impressive.

The company is about to launch its own supercar, named the HIPERCAR, which in a semi-acronymic way stands for 'high-performance carbon reduction' – an absurdly sensible name considering it's styled to look like some form of automotive sci-fi fantasy. Saunders is immediately aware of the strange contradictions inherent in the supercar breed.

'It's a bizarre product. They tend to get driven round relatively slowly with the odd traffic-light Grand Prix, making a lot of noise. We had someone here from America the other day, and he asked where he could go to see lots of interesting cars. And I said that your best bet is probably Mayfair. You'll see everything. A Bugatti; every kind of Lamborghini; all the Ferraris, Bentleys, Rolls-Royces. In this country most supercars are in London. I'm sure [the concentration of ownership] is the same round the world. In the US a lot of them are in Beverly Hills. It needs to look sensational, make the right noise and be fast away from the traffic lights.'

But it's also vitally important that it can perform quickly round the track, at least in theory. It's all about bragging rights, top speeds and lap times.

'As a supercar manufacturer, you are ranked by your performance. We will want our cars tested on a track by journalists and test drivers, whether it's *Autocar* or *Top Gear*. We can't say we're not going to do that: a "we don't do track times" attitude. You'll just get a "couldn't do it" rating.'

Saunders worries that eventually the bubble will burst. People will ultimately realize that there's hardly anywhere that these cars can be used to their full potential. 'Take the Bugatti Chiron. It does, what, 250 mph. But has any Bugatti owner actually done 250 mph? Where would you do it? It's a technical achievement. You have to take your hat off to these people. But people don't use it for that. They use it for driving round town, showing off and having a nice car.'

He's concerned that people will lose interest, especially as many supercars are, he feels, quite boring to drive at low speeds.

'We had a test day a few days ago and sampled a lot of supercars. Sounds like a hard day out! But everyone came back so disappointed. Ferraris and Lamborghinis have become better cars in that they're actually more reliable and easier to drive. Your average Ferrari has been designed, wind-tunnelled, engineered and tested at Fiorano. It'll do a fast lap time at Dunsfold, in the hands of a racing driver or a journalist, but they can be a bit soulless. People say the same about 911s. A new 911 is a fantastically fast car. But getting into an old one – people get excited about that. It's a strange world, the supercar world.'

There's a difficult balance to strike between comfort, raw power and safety. Take the legendary 1992 McLaren F1, which was the most sensational car of that decade and perhaps of all time. A true icon, designed by Peter Stevens and Gordon Murray under the leadership of Ron Dennis; just 106 cars were produced. Ownership of this seminal car, which currently is valued at about £20 million, grants you access to one of the most exclusive automobile clubs in the world, second only to having a Ferrari 250 GTO in your garage.

TV presenter Tiff Needell tells me more. 'It had 600 horse-power, which, with no computers involved, is huge power – as Rowan Atkinson proved. It's got no traction control and 600 bhp is too much for an average driver on public roads.' He's referring to the *Mr Bean* actor-comedian twice crashing his much-loved burgundy model, which on the second occasion needed over £1 million of repairs. That is quite some insurance claim. Despite such misfortunes, he still reportedly sold it for almost $12,000,000 in 2015, substantially more than the sum he paid for it.

Now, supercars have become much more sensible. 'All they're doing is giving you huger and huger headline horse-powers,' says Tiff, 'and then giving you bigger and bigger computers to make sure you never have that power. So, when you open the throttle you don't get 1,500 horsepower, you get about 200. They can't be let loose without traction control or else they'd all be in the ditches. You'd have more fun in a Mazda MX-5 or Caterham 7.'

The supercar grows up

I can recall driving supercars myself for the first time in the 1980s. They'd looked brilliant in magazines and as posters on my bedroom wall, but the reality was often a disappointment. A Lamborghini Countach springs to mind. It was bright red and borrowed for filming a test against the Ferrari Testarossa. The scissor doors were scene-stealingly gorgeous and the noise of the 4.8-litre V12 thrilling, but the crude build and seats reminded me more of a kit car than the apex of automotive engineering. It seemed to require two feet and two hands just to depress the clutch and operate the gear lever. I was thoroughly exhausted by the time I'd arrived on location. The Testarossa was rather better, incidentally, and a foretaste of how supercars would develop. Although wide for its day, it offered impressive visibility, comfortable seats and surprising refinement. You could go for a long drive without fear of nervous exhaustion and actually enjoy being in command of its delightfully musical flat-12 engine.

One man who agrees with me is Dan Parry-Williams, the director of engineering design at McLaren Automotive. This is the sister company of McLaren Racing, one of the world's oldest and most successful Formula One teams. It built on the influence of the McLaren F1 and launched a new series of supercars starting with the 12C in 2011.

'If you go back to the old Lamborghinis they were difficult to get in and out of; you couldn't see where you were going. You certainly couldn't see where you'd been. Broadly speaking they were uncomfortable and intimidating. I don't want to single out Lamborghini but that was the general sense of what you'd expect in a supercar. The cars would get horribly out of shape and scare the living daylights out of you. There was a sense that they weren't accomplished track cars and they were very compromised road cars.'

McLaren's probably done more than any other to make supercars more accessible and enjoyable as a day-to-day driving experience while retaining their competence and speed on the track. To Parry-Williams, the way to avoid a future decline in interest due to the cars being boring at everyday speeds is to engineer in more low-speed tactile feel and involvement.

'We need to jealously guard the characteristics that people like about our cars. Our cars are about driving pleasure and experiencing what [our customers] might call engagement. They're having an exciting drive in the car because we've worked hard on all touch points, whether it's the calibration of the throttle, the brake feel, the steering feel in terms of speed and lateral acceleration build-up, the vibration in the cabin, the noise quality inside the car as well as outside, or the gearshift calibration. All these things are carefully designed and tuned so that the response of the car is proportional to what you're putting in – if you want to drive the car harder your experience is more rewarding as a result. We have to make sure we safeguard that as we move into a new era where maybe we haven't got the traditional soundtrack of a multi-cylinder big engine.'

I first visited their stunning Norman Foster-designed curved glass lakeside headquarters in Woking when it opened back in 2005.Production is now set to expand to 6,000 cars a year and the company has announced that it'll be close to 100 per cent hybrid, with eighteen new models announced before 2025. Only a very few ultimate niche vehicles will be non-hybrid.

Parry-Williams thinks that electrification and using lightweight materials will most probably ensure the supercar's survival in a potentially hostile world. The two technologies will keep those all-important lap times and performance figures improving, while the environmental credentials continue to increase as well.

'We're wholeheartedly embracing hybrid technology. We've been open about our pathway towards electrification and the number of cars that will be electrified in our future lineup. That's very much the core of our strategy.'

As for their lightweight credentials, the company has a new composite-materials centre in Sheffield; and they have tremendous experience in the field, having made over 15,000 so-called MonoCells since 2010. This is the F1-inspired carbon-fibre tub, in which the driver and passenger sit, that forms the primary structure of all McLaren's road cars. They've been developing the technology to be able to produce higher volumes at more affordable prices. This will mean they can make a larger portion of their cars out of the material. They are also trying to simplify the cars, so that they can make them out of fewer parts. For example, if they can replace a roof structure that was formerly made out of twenty parts with just two complex carbon-fibre mouldings, it would be substantially cheaper, better quality, lighter and easier to assemble. They're even pioneering the use of recycled composite materials that can be used in some areas of the cars.

Short circuits – can supercars become fully electric?

The main cloud on this horizon is rapid change in some countries that might make hybrids illegal. This is a problem because the power-to-weight ratio will suffer. Teslas have brilliant 0–60 times of around 2.5 seconds in their fastest models and YouTube is full of videos of them beating Ferrari 458s and Porsche 911 Turbo Ss but they can't sustain them on a track day. The batteries overheat. Weight and the sheer demands of the track make pure electric supercars if not impossible then pretty frustrating for their owners.

'What Tesla have done fundamentally is that they've specified an energy cell and pushed it much harder for very short periods of time.' Nick Carpenter of Delta Motorsport tells me. 'So, your 0–60 run is way beyond the spec-sheet capability of that cell. But what they've understood is that you can push that cell for very limited periods of time, a few seconds. When you floor it in a Model S you get to 60 in 2.5 seconds but you're never going to do more than 4 or 5 seconds before you reach your speed limit and beyond so you ease off.'

If you want something purely electric you have to accept a few knocks to the performance or the staying power. Aston Martin's first all-electric car, the Rapide E, has a top speed of 155 mph and a relatively leisurely 0–60 mph time of just under 4 seconds. The company has fitted a comparatively small 65 kWh battery, presumably to help keep weight down, though this limits the range to around 200 miles. Porsche appears to be prioritising endurance and quick charging as well as headline-grabbing acceleration figures with its first fully electric car, the Taycan.

Croatian company Rimac and Tesla are throwing bigger and bigger batteries at the problem. Rimac's powertrain is included in the £2 million Pininfarina Battista, which debuted at the Geneva Show in 2019, promising sub-2-second 0–62 mph times, a 217 mph top speed and 1,900 horsepower from its 120 kWh battery pack. Tesla's Roadster 2.0 is set to have a 200 kWh battery and offer similar acceleration and a top speed of over 250 mph, with a range of over 600 miles at highway speeds. Rimac claim that their 1,914 horsepower C Two, with slightly more power than the Battista, will lap the Nürburgring twice without any loss of speed. Whether these cars really will deliver this stunning performance lap after lap without overheating is yet to be proven.

At Ariel, Simon Saunders hopes the politicians won't demand pure electric for the sake of it. His HIPERCAR uses inboard electric motors with a range-extending turbine engine charging the battery. The engine is used only to top up the battery, rather like the petrol engine on the original BMW i3, but is vastly more powerful. It boasts 1,180 bhp, can achieve 0–60 mph acceleration in 2.4 seconds and 0–100 mph in 3.8 seconds. It will offer low levels of pollution and high levels of economy, and the ability to keep driving all day if you want to.

'A turbine which will be running at 120,000 rpm is very, very efficient once it's up to speed,' says Saunders. 'And, once it gets going, it's almost emission free. Delta Motorsport are working with Johnson Matthey to make the start-up procedure as clean as possible. There's a certain amount of emissions in the start-up but not much. The range extender is very, very, very low emissions. Our calculations show that we can go for 500 miles on a 30-litre tank of fuel.'

Ripping up the supercar rulebook – Ariel's gas-turbine-electric 1180 horsepower HIPERCAR.

Turbine cars are nothing new of course. Rover built a prototype called JET1 in 1950 that went on to record 152 mph over a flying kilometre at Jabbeke in Belgium in 1952. Chrysler lent a limited run of fifty experimental cars to potential customers in the 1960s. They offered a spectacular soundtrack but proved thirsty and suffered from terrible lag on acceleration as the turbine gained speed. Their exhausts also tended to singe pedestrians' legs. Making the turbine part of a hybrid powertrain means it can operate at maximum efficiency, helping to keep heat within the system, while battery power can be used to compensate for any lag.

A turbine is also much lighter than batteries would be to achieve the equivalent power. 'Added lightness' was the 1950s catchphrase of the remarkable Colin Chapman, who invented the rear-engine racing car and produced remarkable classics like the original Lotus Elan and Lotus Seven. 'It's more true than ever today,' says Saunders. 'It's not necessarily the cheapest way, but certainly the easiest way to improve performance, fuel economy and so on is to make vehicles lighter.' The company has even experimented with creating titanium components – a monumental headache, as titanium can be engineered only in a chemically inactive atmosphere. This can involve an airtight tent full of argon with people reaching in through sealed sleeves; most annoying if a tool needs to be retrieved from outside, as the tent would need to be purged of oxygen once more.

Certainly as it stands, zero emissions for the sake of it may well compromise the sheer unassailable stamina that is the raison d'être of supercars. And then we'd have to make do with survivors from what will be seen as a golden age of performance road cars. Whatever happens to the power source, though, changes elsewhere in the car industry will definitely have their effect at the fast end too.

Supercars and future tech

With supercars so eye-wateringly expensive, Parry-Williams at McLaren Automotive thinks we could see a move to car sharing rather than ownership. 'You won't necessarily have to own the vehicle to get your hands on one. It'll show up when you want it and you won't have to worry about it when you're not using it.'

Professor Dale Harrow from the Royal College of Art sees the potential for supercar ownership of the future to go in two different directions. On the one hand, a company might produce a series of cars that are used exclusively on race tracks – so, more like owning a luxury yacht or a racehorse. On the other hand, it might create a whole service experience with hotels and resorts where the car can be driven in a unique environment.

McLaren is already offering its wealthy customers a broad range of holidays and other adventures, whether driving through the Italian mountains or on historic race tracks like Hockenheim. As of October 2018, its website featured this enticing invitation: 'Winter may be fast approaching, but our programme of events isn't cooling off just yet. Whether you put yourself in the hot seat at the legendary Kyalami circuit as part of our Epic 9-day driving tour across South Africa, or join us 200 miles north of the Arctic Circle for an exhilarating experience on the ice – ensure your winter is truly epic.'

The supercar club is a thriving international phenomenon that looks set to flourish. You pay a subscription and indulge in events where you can take cars to the limit. If (like most of us) you can't afford to buy your own, you can share one of the club's garage of exotica. Britain's Auto Vivendi has a fleet that includes a rotating selection of Lamborghinis, Ferraris, Aston Martins and McLarens. Membership fees to drive them are between £12,500 and £33,500 per year – which doesn't seem so extortionate when you consider that a major service for a Lamborghini costs around £4,000 and a set of tyres about £3,000.

Freedom Supercars in the US offers a similar deal but the really exclusive clubs, like the Supercar Owners Circle, set up by Swiss brothers Florian and Stefan Lemberger, require you to own your car. Membership is restricted to those with a collection considered worthy, probably valued at more than £5 million. You get access to bespoke track days, fast-track passes to the latest cars from select manufacturers and excursions like Alpine trips over specially closed roads. It's a sort of Davos on Tour for the world's elite car owners.

Brands will increasingly broaden out to offer more than just the car itself; ownership appeal will be as much about shared experience as it is performance. This will be especially true in the supercar sector. Aston Martin are also keen to offer hotels,

Merch too far?

How far can you extend a brand? Ferrari knows no shame, it seems. Its first branded amusement park opened in Abu Dhabi in 2010, with Portaventura following in 2017. More are planned for Spain, China and the US. Then there's the Ferrari-branded Segway that, at $10,000, was nearly twice the price of the original and made you look twice as ridiculous. The bright-red branded Acer netbook wasn't much better. It was one of the slower computers you could buy at the time, even with the vroom sound effect on start-up.

Overpriced and poorly specified mountain bikes are another speciality with luxury car brands but there are also plenty of exceptions. The 2011 £12,000 Specialized S-Works McLaren Venge justifiably laid claim to being the fastest road bike in the world at the time and the 2012 £25,000 Aston Martin One-77, with its sophisticated on-board computer, body measurement and touch-screen, was widely regarded by pundits as the most advanced bike in the world when launched.

Perhaps the ultimate brand extension is the merchandise associated with the £2.5 million Bugatti Chiron. I'm not talking about the £3,000 leather jacket, or the £400,000 Jacob & Co watch, but the Niniette 66 yacht, priced at $4m and including a hot tub, a fire pit and a champagne bar. Buyers beware however. There is rumoured to be a longer 80-foot model on the drawing board, and you wouldn't want to be trumped.

yachts and other luxury experiences. It makes a welcome change from most off-topic brand extensions by car companies, which have long been rather poor quality. But for these more ambitious augmentations to work companies will have to take them very seriously.

Autonomy is a development that will mix very well with supercars and speed. Many manufacturers are already considering the inclusion of in-car expert electronic driving tutors – something like KITT in *Knight Rider*. Heads-up displays will communicate with the driver on how to corner, suggesting lines through a bend; like a video game, but for real. The idea that a virtual Lewis Hamilton could teach you how to drive on a race track is going to be really compelling for the person who owns that Mercedes.

Porsche CEO Oliver Blume is on record as looking forward to the day when you can compare your own driving abilities with a Formula One driver. If you find out you're ten seconds slower than Mark Webber round a track, the onboard system would analyse why by looking at your braking, racing lines, etc. It's a mixed- or augmented-reality driving format, with information added in your field of view as a driver. This information will be filtered, so you receive only what you need, and on both road and track it will assist with driver skills, keep you out of trouble and help you get the most from the car while keeping you safe.

The high cost of supercars increases the likelihood of their manufacturers having the margins to introduce new assisted-driving technologies. Saunders suggested that high-end manufacturers might develop the new technology, get the patents and then sell it to the volume manufacturer, who would later use it in such high volumes that the cost would go down. Volkswagen have already shown a virtual race trainer based on a Golf. You can use it to get a handicap for your driving talent: as in the sport of golf, you get a par for the course, here set by an AI driver, which you have to try to match or surpass.

Racing

The visceral draw of speed is not only for the driver. From the car's early days, speed has also been savoured by spectators vicariously revelling in the thrill and danger of competition. Across the world, racing has created its own hallowed temples of speed, from Bonneville and Brooklands to Indianapolis and the Nürburgring. Race-car technology has also helped to develop more quotidian road models. As the technological demands of motorsport depend on efficiency and constant progress, developments in tyre technology, braking, hybridisation and aerodynamics often first appear on the track.

It's certainly a big industry. Formula One, traditionally the pinnacle of the sport, was sold to Liberty Media in 2017 for $4.6 billion and claims over 490 million viewers worldwide. And the taste for watching cars go round and round in circles is being particularly keenly adopted by emerging economies. Brazil is F1's biggest market, with 115 million viewers, and new circuits like Hanoi in Vietnam are part of an expanded calendar. In

Britain alone the industry is estimated to be worth £10 billion to the economy.

Will the industry go through the same upheavals that the road-car industry is currently experiencing? How will it be affected by electrification, autonomy and the explosion of environmental concerns? For some, the future survival of the sport depends upon embracing electrification. And this is on a total basis, not just the hybridisation seen in Formula One's energy-recovery systems. But if electric supercars are difficult to produce, the demands of track racing are even greater. A spike of power for road use lasts a few seconds but in racing it is sustained for lap after lap after lap. Batteries drain incredibly quickly and overheat fast. If you need ten times the power, you've got ten times the current. As the heat generated is proportional to the square of the current, you've then got 100 times the heat. If you don't want everything to melt, thermal management is a big issue.

In spite of this, Formula E, the all-electric racing formula operated by the world motorsport governing body, the Fédération Internationale de l'Automobile (FIA), is leading the way. Conceived in 2011 and with the first championship starting in Beijing in September 2014, it claims it's 'reinventing racing for the 21st century'. It's cunningly devised. They've gone for city race tracks, like Paris, Hong Kong, Monaco and New York, because that's potentially the biggest market for electric cars. While working hard to ensure that the 335 bhp cars have attractive styling, they've kept the bodywork and chassis stock so they can stop teams spending a fortune fiddling about with aerodynamics and the suspension. They've also compelled all teams to use the same battery. This stops any one of the teams doing an exclusive deal with whoever's battery technology is currently judged to be best.

Consequently, by international motorsport standards, it's relatively cheap. Formula E costs about $40 million per team per year, which is a fifth of the cost of Formula One, but it already attracts half the global audience. Not surprisingly this is proving an appealing option for car manufacturers keen to establish their electric credentials. Formula E's second-generation car appeared for the 2018–19 season and is raced by manufacturers including Mercedes (with Venturi), Audi, Porsche, BMW, Jaguar, Nissan, DS-Citroën (with Techeetah) and Mahindra as well as new Chinese EV manufacturer NIO. All this manufacturer interest in Formula E will undoubtedly enhance the image of electric vehicles.

Phil Charles is the technical manager of the Jaguar Formula E team. The team operates from the Williams Formula 1 team's headquarters near Didcot in Oxfordshire. The building contains the largest private collection of Formula One cars in the world, all squeaky clean, polished to perfection and dramatically displayed. I strolled round them with Charles, who studied automotive design at Loughborough prior to specialising in vehicle dynamics and aerodynamics. He worked with the Renault F1 team and Toro Rosso before taking up the Formula E challenge.

The cars in the museum represent an era where the most competitive developments were in aerodynamics – bodywork tweaks to smooth airflow round brake ducts or enhance ground force, for example. With Formula E, the competition is all about how to manage the available energy from the supplied battery through the motor, the inverter and the gearbox – the bits they've allowed teams to play with. The progress has been so quick that the championship has advanced from drivers having to change to a second fully charged car during the 45-minute race to one where a single car can keep going for almost the same duration.

'In Formula One I was employing people who were vehicle dynamicists and aerodynamicists,' says Charles. 'Here I'm employing mathematicians. Because they're doing a lot of work on the energy side – to be really clever with the energy.' The result seems to be that drivers are in constant communication with the wider team during races. One aspect of this driving teamwork is 'energy cycling', or how to conserve energy round a lap. The power has to be 'scheduled' to eke out its full potential.

'In the Formula E car we tend to accelerate, accelerate and accelerate and then get to a certain point on the straight where we lift and coast so you cut the power and pull the regen paddle. That point is particular to how you do that straight in the fastest way you can with a given amount of energy. We work all this out. We give Mitch [Evans, one of their drivers] lights and beeps and some of it is automated for him. We can't do it live for him but the car is programmed to help him.'

It's the same on the bends. 'Let's say, for example, that for a low-speed corner we make the car balanced so that it rotates itself. He turns the steering and the car turns easily. If it understeers and he has to use the throttle to turn the car, then he's burning energy to turn the car. It's simple but the way we balance the car around him helps him save energy. That's kind of key. A tyre slide somewhere along the line will cost you energy. The maths guys we are using to look at the energy cycle are also looking at how Mitch is driving all the time.'

I wondered if this all made the cars more boring to drive. I spoke to Mitch Evans, who is from New Zealand, started with GP3 and who was, at sixteen, the youngest person ever to win a Grand Prix. He confirmed that the required driving style was rather different, and that he found it more difficult to detect wheel spin in an electric powertrain, but if he had any complaints he wasn't admitting to them.

TV presenter Tiff Needell was less equivocal. 'The throttle response isn't so good. Though you have all this peak torque from zero, you're stuck with one level of torque whatever your throttle position. People rave about the torque but you don't want a lot at times because it breaks traction. So unless you have traction control you get very embarrassed very easily.'

Whatever the truth about the driving experience, it is certainly a fact that electric racing is continuing the trend – which is occurring in conventional racing too – that the driver is becoming more of an operator, rather than solely responsible for their own decisions.

Another concern I had at Didcot was whether the model of motorsport using closely controlled formulae could prevent useful innovation, both for the sport and the much-vaunted crossover between racing cars and road cars. It tends to favour incremental optimisation and little tweaks. Greater freedom in what the teams can do might yield more interesting and radical progress. Formula E, for example, demands a single rear-mounted motor driving the rear wheels through a gearbox. It therefore rules out experiments with developments like motors in wheels and the more sophisticated approach to torque vectoring this allows.

Torque vectoring refers to technology that can vary the amount of torque sent by a power unit to each individual wheel. This can be used to help a car corner, accelerate and brake more effectively. With a central battery and a power unit in each wheel, designers can control the power distribution by computer, substituting a cable for heavy driveshafts and gearboxes. There would be a huge incentive for Formula E teams to develop this technology, if only the Formula were less prescriptive. Phil looks forward to a more varied future when such things are possible, maybe with the next generation of the car in 2025.

Finally, there is the elephant in the Formula E room; the most obvious question for someone coming from Formula One to Formula E, and a key one for the future of electric motorsport: what about the lack of noise? For Phil Charles (rather predictably) it is not a major issue. 'When you go to a Formula One garage and hear the noise, that hits you in the face. But when you're tuning in on TV you're not getting that aspect. I want to see the [Formula E] championship maintain really good racing. Because if you're able to overtake it's really entertaining. And when you actually get people there and they see that the cars actually look pretty quick, they're going round tight and twisty circuits, they're overtaking each other, and they have crashes and dings as they're trying to do it, then the noise gets forgotten because it's so exciting anyway. Formula One noise isn't pretty – it's more of a kick-in-the-nuts-type noise, and personally I don't miss it.'

Electric sprints

While endurance racing and all but the shortest of rallies look like nonstarters for electric power as it stands, Formula E's success has inspired suggestions for other forms of electric motorsport, particularly those that require short, powerful sprints. These include rallycross, hill climbing and, most intriguingly, the land speed record.

The current holder of the outright World Land Speed Record is Thrust SSC, a twin turbofan jet-powered car that achieved 763.035 mph – 1227.985 km/h – over one mile in October 1997 driven by RAF pilot Andy Green. This was the first supersonic record with the car cum rocket reaching a scorching Mach 1.016. Its successor has been relaunched, under new management, as Bloodhound SR, and is aiming for 1,000 mph.

There are many and varied land speed records and electric cars are already holders of some, but they're some way off the outright pace. The current electric land speed record of 341.4 mph was set by the Venturi Buckeye Bullet 3 (VBB-3) on the

Esports

Removing the noise from motorsport is one thing, but what about removing the whole car? Esports – multiplayer video games played competitively for the benefit of spectators – will become ever more closely entwined with motorsport. Though slow to take off compared to first-person shooters, battle-arena and strategy games, race-car esport is now becoming an ever more important part of what, by 2022, should be a $2 billion industry with over 600 million viewers.

No other esport so closely parallels the real thing, and in some ways it's better. With no risk of personal injury and shattering priceless carbon fibre, the racing can be closer and riskier – and driving skills can be honed over hours of virtual practice. The drivers aren't hidden in their cockpits and you get to see them hard at work. In the future, sponsors may even defect to esports events to reach a younger audience.

Whatever your preferred game, there's an esports championship to go with it, including Gran Turismo Sport, Forza Motorsport and rFactor. Meanwhile real race formulae are branching out into esports, from NASCAR to Formula One. F1 esports has four qualifying rounds, organised in tandem with real races, culminating in a stage where the esports F1 teams – versions of real ones – pick their drivers for the final. The winner of F1 esports gained a real drive with Mercedes for the first time in 2018, and a cut of the $200,000 prize fund.

With the colossal cost of entering real motorsport a major handicap to discovering future Ayrton Sennas and Jack Brabhams, esports could solve a talent shortage as the new grassroots of motorsport. You'll progress from one to the other. You'll just need the price of a game (and usually steering wheels and pedals) to take part rather than the usual arm, leg or kidney.

Feet up and flat out as the 2018 F1 Esports series gets underway at the Gfinity Arena in London.

Bonneville Speedway on 19 September 2016. Developed by a team of students from Ohio University, it is four-wheel drive and highly streamlined with an extremely low drag coefficient. The car uses two motors and can vary its power split between front and rear axles, usually favouring the front to avoid 300 mph powerslides. The main problem for electric land speed record cars is finding somewhere to run them. Thrust SSC strutted its stuff in a South African desert which provides plenty of room, while the sandy surface isn't a problem for its jet power source. But electric cars are dependent on putting their power down through rubber tyres and they need a higher friction surface. Amazingly this, more than anything else, may stop electric land speed record cars reaching the very highest levels of performance.

Rallycross may embrace electric motors in the relatively near future. It's long been one of the most spectacular and spectator-friendly motorsports. You can see the whole course from one vantage point, the races are short, overtaking is constant and there's plenty of body contact. I can vividly remember watching races broadcast by the BBC from Lydden Hill as a child in the 1960s with somersaulting Mini Coopers and Hillman Imps battling it out with Escorts and Cortinas. The top cars are now four-wheel drive. The new four-wheel-drive electric ones on the drawing board feature two 250 kW electric motors, which together will produce the equivalent of 670 bhp, powered by a 52.65 kWh battery. The drivetrain is built by the Austrian company Kreisel and can be retrofitted into the existing cars' steel bodies. Volkswagen, Audi and Peugeot are said to be keen to support the sport's move to electric propulsion and were aiming to introduce their electric cars to run alongside conventional internal combustion engine World Rallycross Supercars for the 2020 season. Not all manufacturers are keen to embrace the switch, however, and the move has been postponed to 2022. A

Frenchman Romain Dumas smashes all records up the epic Pikes Peak Hillclimb in his all-electric Volkswagen I.D. R.

one-make junior version of the championship, though, is still scheduled to go ahead in 2021, using 250 kW four-wheel-drive cars made by the Spanish firm QEV Technologies.

For inspiration and a reminder of how electric cars can lead the way, one needs to look to another field of motorsport where they are proving their prowess and popularity: hill climbing. Pikes Peak is a brilliantly mad event that takes place every year on the 14,115-foot mountain in the Rockies, North America. A twelve-mile-long road with sheer drops climbs right to the top. It's a thrilling and dangerous course for a hill climb, even though it's now fully tarmacked and the cars go one at a time rather than in groups.

In 2018, Volkswagen proved electric propulsion's potential when double Le Mans winner and former World Endurance

Champion Romain Dumas raced the company's all-electric I.D. R up the hill, not only smashing the record for electric cars but also besting Sebastien Loeb's long-standing record, set in 2013 in a Peugeot 208 T16. Loeb's time was 8:13.878. Dumas and the VW I.D. R finished the course in 7:57.148, despite a track that was slightly damp in places. An advantage of electric cars is that whereas combustion engines lose performance as they climb and the air thins, electric motors are consistent in their power output whatever the altitude. They can also exploit the fact that the thinning air offers less aerodynamic resistance.

Unlike Formula E or Rallycross, the hill climb is lightly regulated – and VW could let loose their design creativity. They maximised downforce through aerodynamics and took advantage of the relatively short course and relatively low speeds by keeping the batteries in the 670 bhp car small to save weight. They also used carefully designed air flow to cool them. It's now thought that it will be hard for a combustion-engined car to beat VW's record in the climb, such was their expertise in balancing the car's aerodynamics, cooling flow, battery size, weight distribution and power delivery. VW even compensated for the quietness of the car by installing a hooter.

Autonomous racers

Some people see not only a future for electric power in racing but a role for autonomy too. Roborace is a company that was founded by Russian smartphone entrepreneur Denis Sverdlov in 2015. It specialises in developing autonomous racing cars. I met its strategy officer, Bryan Balcombe, at the Goodwood Festival of Speed, where his company's Robocar became the first-ever fully autonomous car to ascend the venue's famous hill climb. He had the studied, calm good nature and open smile of a man who's relaxed in the face of often hostile criticism. His background is conventional motorsport and he's served sixteen years in Formula One as a systems engineer.

The Robocar was designed by Daniel Simon, who also created the vehicles for the film *Tron: Legacy*. It's about the same size as an F1 car, has a 540 kWh battery and four separate 135 kW motors. Costing around £1 million to build, it has a claimed maximum speed of 195 mph. Its autonomous technology is similar to those of its road-driving cousins and the intention is to set up a 'driver's championship'. As Bryan explains, 'We have radar, LiDAR, sonar, machine-vision cameras, GPS and inertia sensors and slip-angle sensors [as well as] all the normal motorsport sensors that you'd have on a car such as tyre temperatures, tyre pressures, suspension movement and everything to do with the powertrain. So, all of the data is provided to the AI driving software and the teams work on providing the best AI driver.'

The traditional motorsport influence is evident in the standardised elements of the car. 'We run a standardised hardware platform so all the vehicles are identical. At the moment all the sensing and communications are identical and the teams just write the software. In terms of their freedom they can write whatever software they want in whatever way they want. They just have to pass a driver-licensing test before they're able to compete.'

Bryan envisages teams of autonomous cars competing against each other. 'All the sensor data is identical so you have a fair starting point. What sensors do you choose to use? How do you weight those? What computer algorithms do you have attached to them? Those are the decisions that need to be made for you to perceive your environment. That's the challenge number one. Challenge two is once you know where you are in the world and what the other objects are, what's your driving or racing line that you're going to take? Your trajectory. That's a really challenging task in itself.

'That's really how we point the competition. Perception is one level. And above that is the driving task, which is what as humans we want to see. Racing interaction: two cars at least interacting with each other.'

The cars learn the tracks where they race in slow motion and have to pass a qualification process in what they call a DevBot vehicle, a version of the race car that can also be human-controlled. 'We need to do that from a safety perspective and also from an entertainment and performance perspective. We want to know that these AI drivers are good-quality drivers.

'At the moment we find the AI software is 10 per cent or 20 per cent below where a human driver is. Most of that is because of the safety margin, which in our public demonstrations we've been willing to take. We leave a 1-metre safety margin to the wall at Formula E tracks, whereas a driver will nudge up against the wall if they're on the limit. We don't ride the kerbs when we go through chicanes, but a driver will and there's a lot of time to be made in those areas.'

Goodwood provided some AI challenges. The car is usually raced at Formula E events where the track has concrete walls either side. The car can detect where it is in relation to vertical surfaces rather more easily than the edges of the Goodwood track, which are grass.

'When you're trying to use LiDAR on a horizontal surface it's difficult to differentiate because there's no height variation. So the only way to do that is looking at machine vision and looking at the difference between road surface and grass, which is called image segmentation.'

Fortunately they'd already done work with grass using hyper-spectral cameras that are now being used in agriculture to look at crops. They use infrared and other ranges of light beyond human vision and can take account of the way grass changes colour, height and surface characteristics depending upon the time of year and the weather conditions. But, even so, I remain sceptical.

I think it's hard to envisage this form of autonomous motor-sport having much appeal unless it becomes a sort of *Robot Wars*-style crashfest and I can't see the serious governing bodies of motorsport sanctioning that. But Bryan was obviously not perturbed by regulatory hurdles and even had another, rather more extreme-sounding suggestion that involved mixing autonomous or highly assisted race cars with actual road traffic and encouraging no-holds-barred racing, all on real roads, among everyday cars, buses and trucks, driven by the likes of you and me.

'If motorsport's going to play a role in developing Level 5 autonomy for the road – vehicles that can go anywhere; any situations, any roads – that technology has got to grow beyond the race track. It's got to start looking at other types

Vrrrrrrrr!

Formula One may boast the biggest global audience of any motor sport but there's a constant worry that not enough young spectators are attracted to the sport. It's a problem usually blamed on racing being too boring. Cars go round in processionary circles and the circuits conspire to prevent overtaking. Far from gladiatorial combat, miniscule iterative engineering provides the edge. Sometimes the most interesting thing about the sport seems to be the politics.

Maybe the answer lies with technology, including virtual and augmented reality. Youngsters who aren't attracted to the sport could be put behind the wheel feeling the forces, the vibration and the noise through a combination of headsets, exoskeletons and haptic smart clothing. Sensors on the driver would transmit the sensations they feel to the audience. Spectators would be able to see the instrumentation, touch the controls and sense the airflow round the car. Even the driver's brain signals could be read and communicated to viewers at home, who would also be able to customise their cars, get a fully immersive view of the race and have the option of competing against the real drivers.

of formats, challenges and competitions that really develop the software that's required. That would mean looking at real road environments filled with traffic and having cars racing through that traffic environment. A completely different type of motorsport than we've ever seen before.'

In the meantime, driver assistance is already happening in some branches of the sport. Bryan tells me that 'assisted driving is almost creeping into Le Mans at the moment. In the GT class of racing you have a radar system and a camera system, looking rearward and automatically identifying the cars that are coming past you with a range of about 250 metres, giving the driver information about when to let a car past that isn't a competitor. In a way that's an assistance technology like an advanced rear-view mirror, if you like. It doesn't control the vehicle. If it started to control it we'd say that was augmented. Then you have both the human and the machine trying to control the output of the car.'

Future racing technology will be about more than electrification and autonomy. There's a whole host of other innovations that will smarten up the sport. Augmented and virtual reality will link audience and driver to provide new levels of spectator involvement. The use of more advanced composites and new wonder materials like graphene and carbon nanotubes will also open new avenues to explore. Bodywork will change its shape on command and self-repair. Tyres will morph too. Hankook has shown a prototype tyre that broadens out on corners, exposing an aerodynamic insert that boosts ground force but then slims down on the straights for lower wind resistance.

Others reject any suggestion of a techno-utopian view of motorsport. Maybe we should just revel in the petrol-head gloriousness of traditional motorsport as the only place in the world you can still go to hear the sounds, smells and raw competition of petrol engines thundering round a track – an antidote to the boredom of an autonomous and electric world. That's certainly Tiff Needell's view.

'Motorsport will eventually realise it can't be the science lab anymore because science has overtaken us. The more the world goes autonomous and electric, as it inevitably will, the more motorsport will be petrol and entertainment and hopefully we'll just have wild motorsport for things you never see on the roads. Like NASCAR. They've got 300 quarter-mile dirt ovals with racers going round every night of the week. In American motorsport they're still using 1950s cars pretty much. No driver aids. Just big, rorty V8s howling round in any sort of big old chassis you want. We'll be like the gladiators of Roman days – entertainment that's enjoyable for both driver and spectator.'

There's no room for autonomy in Tiff's view of the world. Nor for that matter in Ross Brawn's. He's the new boss of Liberty's Formula One empire who's revising the championship in 2021 with entertainment in mind. Quoted at the Ferrari exhibition at the Design Museum, London, in 2018, he said:

'I think the change in transportation is a fascinating area. We are challenged in Formula One as to how much relevance we should have and we believe our fans and our enthusiasts enjoy a certain aspect of Formula One, and we can't necessarily maintain that or create it if we go an electric-car route, or even

So much more than a black circle – Hankook's prototype racing tyre that maximises downforce and minimises air resistance.

an autonomous-car route. The last thing we need is autonomous cars in Formula One, because the driver is the gladiator, the driver is the hero, and the one we want for people to really engage with.'

But might this petrol-head's dream be impossible because of an environmental backlash against traditional motor sport – with all those resources devoted to going nowhere, all that degradation in air quality?

Fuelling waste or driving innovation?

Racing cars are typically designed to be fast, not to meet emissions and environmental regulations. The smell of burning race fuel has often been quoted as part of the appeal of motorsport events but will we soon start worrying about breathing it in? Then there's all the tyre and brake dust. At the Goodwood Festival of Speed in 2018, I even found myself worrying for the first time in my life about what I was inhaling while watching some drift cars shredding their tyres in the cause of entertainment. Attitudes may be changing without us old guard realising it.

Evidence of direct pollution is limited because, for some reason, people don't seem to put up air-quality measuring stations near motorsport venues. However, it's certainly true that Formula One cars used to be gas-guzzlers. Back in 2007, each race car emitted around 1.5 kg of CO_2 for every kilometre it drove, or about nine times that of a family car, but they've since become more economical. Race fuel has become more like standard petrol. Early fuels were reputedly so potent that the engines had to be washed in ordinary petrol at the end of the race to stop the fuel from corroding them. They contained a huge array of chemicals and additives, featuring large quantities of benzene, alcohol and aviation fuel.

Studies of a race in Montreal noted that race days tended to coincide with high pollution levels, particularly if it was a day when there was a thermal inversion. But most of the nasties were the result of general traffic and fireworks displays rather than the race itself. And there was another vital statistic noted: that one private jet flying in for the race emitted more toxins than all of the race cars *combined with* the emissions from the cars of all the spectators travelling to the event, which rather puts things into perspective.

A study of the Grand Prix of Baltimore – an IndyCar Series race held on a street circuit in Baltimore, Maryland, between 2011 and 2013 – concluded that the racing had very little environmental impact and race days had less pollution than usual because more roads were closed, reducing the normal social, commuter and commercial traffic. It's certainly true that the pollution caused by motorsport is almost entirely down to teams lugging their equipment round the world and fans travelling to the circuits. The actual racing constitutes just 0.3 per cent of Formula One's emissions while teams do 100,000 miles a year in planes to test cars and compete. But, then again, air miles are a consequence of many sports, not just motorsport – and indeed of any gathering that brings people in from all over the world, even committees on climate change. Ironically, cycle racing has a huge environmental effect due to the travelling demands of teams, equipment, spectators and the media, even though the racers themselves are obviously, in a narrow sense, zero emission. Also, with cycling there's not even the corollary benefit that technologies on the track can yield improvements on the road.

Greenpeace has previously tried to disrupt Formula One events in its habitual way – climbing on buildings with banners and suchlike. But the anger has normally been directed against corporate sponsors – for example Shell and their Arctic drilling plans – rather than the sport itself. In fact the charity has made very encouraging comments about Formula One. Its former executive director Kumi Naidoo was an ardent devotee, praising its 'technology, sportsmanship and innovation'. Individual teams have also tried hard to become more environmentally conscious. McLaren for example became carbon neutral in 2011, largely through controlling its emissions but also with some amount of offsetting. It recycles two thirds of what it throws away and sends zero waste to landfill.

So, motorsport looks like it's going to generate no more environmental angst than any other global sport with roving teams and mass audiences. However, to keep appealing to a wide audience, it is going to need to embrace both its noisy petrol-head traditions and an exciting technology-packed future. The gladiatorial glories of the past will live on while technology, from electrification and esports to radical design and virtual reality, will make it more immediate to a wider audience while re-establishing the positive link between race-car and road-car development. Once again, bleeding-edge development on the track will drive progress on the street.

FIVE

HACKERS AND CRASH-TEST DUMMIES

Safety and the Age of Automation

If you ever find yourself in a procrastination cycle watching YouTube, it won't be long before you encounter a 'Bad Drivers' compilation. Whether it's percussive parking, road rage or dysfunctional driving, these videos make clear that for every careful driving instructor there's a plethora of road users who should never be allowed behind the wheel. Road crashes now kill 1.35 million people a year globally. And, despite driving authorities and the industry working hard to improve road safety, particularly in more developed countries, the sheer success of the car has meant that this annual global death toll has continued to rise.

The first recorded fatality caused by an automobile was in Ireland, in 1869, when Mary Ward was thrown from an

Ugly ducklings

The 1950s saw a bizarre collection of concept cars designed in the US to signal a new era of safe driving. What they lacked in practically, they often made up for in their monstrosity. The most spectacular was the 1957 Aurora, built by Father Alfred A. Juliano, a Catholic priest. The nose was designed to be a padded scoop to catch errant pedestrians and provide crush space. The windscreen swelled out miles in front of the driver so their head didn't hit it in a collision. There were side-impact bars, a built-in roll cage, seat belts and a padded instrument panel.

Slightly more conventionally shaped but no less innovative is the 1957 safety car conceived by Cornell University and insurance company Liberty Mutual. Seat belts, nets to help prevent whiplash and extensive padding were all features that would be adopted by later production cars. More unusually there was no steering wheel to impale the driver but two sliding levers surrounded by cushioning material.

Frustrated car designer Father Alfred Juliano wanted to build the world's safest car, but instead his Aurora achieved global automotive fame for its sheer ugliness.

experimental steam car and fell under its wheels. In 1895, there were only two cars in the US state of Ohio, yet they still reportedly managed to run into each other. So far this century, more than 650,000 people have been killed on the roads in the US alone. Sadly, when humans and cars meet, bad things can happen. Consequently, from the outset, the evolution of the car has moved in step with technology designed to prevent its devastation.

Safety statistics must always be interpreted with care, but deaths and serious injuries have often trended downwards in wealthier countries. The UK, for example, currently has some of the safest roads in the world. Measured by road deaths per 100,000 population, it was behind only Micronesia and Sweden according to the World Health Organisation in 2015. I sometimes wonder if this is because the traffic is becoming so bad it's just getting hard to reach potentially fatal speeds. But the research says otherwise.

Cars have unquestionably become safer but it's been a slow process, punctuated by many odd inventions along the way. You can trace the history by browsing the plentiful range of car-crash YouTube videos, as referred to above. Crashes are often the best way of testing safety features and quite coincidentally seem to be endlessly popular as well. When I started producing *Fifth Gear*, one of Channel 5's demands was that we incorporated plenty of crashes into the show. This apparently wasn't just for the wanton destruction; they had to be car crashes that illustrated something in particular. We worked hard to come up with various scenarios: big cars into large cars; new cars into old cars; truck sideswipes – a problem on British motorways, where left-hand-drive trucks change lane and collide with cars in their blind spot. We also investigated the dangers of unrestrained boot luggage, along with killer road signs, kerbs and lampposts that might squash your car or catapult it on a spectacularly dangerous trajectory. We used remote car radio control but with full-sized cars, all developed by an enthusiastic team at Cranfield University.

The allure of twisted metal can almost border on fetishism, as demonstrated by movies such as *Crash* or those odd sixties American rock songs like 'Dead Man's Curve' with macabre themes derived from real-life collisions. Then there's the mystique that surrounds celebrity car deaths. Albert Camus was famously killed in his publisher Michel Gallimard's Facel Vega in 1960, when it hit a tree in Villeblevin, France. Isadora Duncan's silk scarf entangled itself around the axle and open-spoked wheels of her mechanic lover's Amilcar CGSS near Nice on the French Riviera in 1927. She was pulled from the car and broke her neck. Marc Bolan's fabled demise was in a Mini 1275 GT, driven by American singer Gloria Jones, which hurtled into a tree on Barnes Common in 1977 and killed him instantly.

And yet, beyond all the mayhem and morbid fascination, car crashes can be incredibly useful. The car industry has learned from them and safety has vastly improved as a result. First there was the emphasis on structural integrity: if your car could survive being chucked off a cliff and drive off again this was deemed excellent. Then seat belts appeared and brave stunt drivers could perform repeated barrel rolls in what would now be regarded as 1950s classics. The next developments were crumple zones and safety cages; cars became tougher in the middle to protect the occupants, and softer at the extremities to cushion the initial force of an impact. Prototype and production-car crash tests have become a staple for upholding safety standards.

Legislation will continue to get tougher but a big incentive for improvement comes from consumers themselves – people will continue to demand cars that are safer. The importance of the market can be traced back to the work of American consumer campaigner Ralph Nader and his infamous book *Unsafe at Any Speed*. Here he criticised cars like the Chevrolet Corvair with its swing axle rear suspension and rear-mounted engine that made the car spin with remarkably little provocation. He exposed cars with interiors full of sharp edges into which occupants were impaled during impacts – somewhat like mediaeval battle implements. Thanks to his work the US became the world leader in automotive safety. A framework of federal crash-test standards followed and the National Highway Safety Bureau announced its Experimental Safety Vehicle (ESV) project, to encourage the development of safer cars by 1980.

The initial results weren't necessarily attractive, mainly thanks to their enthusiastic use of rubber. One thinks of Volvos with their chunky bumpers or the Fiat ESV 1500 prototype of 1973, a miniature car with massive black growths disguising the bodywork. Joan Claybrook, a former colleague of Ralph Nader, began independent crash tests of cars on the US market in 1979.

They don't make them like they used to. Euro NCAP tests show you'd be killed or seriously injured in a 1997 Rover 100 (left) but likely to get away with minor bruises in a 2017 Honda Jazz (right) in the same 40 mph smash.

Called the New Car Assessment Programme (NCAP), it was the first-ever consumer safety rating system for cars. It harnessed the power of consumer information to accelerate safety. The programme generated instant media interest and manufacturers soon started to compete to make their cars safer. They'd proved the power of the market could improve crashworthiness standards.

In contrast to America, European countries initially relied on more forgiving crash tests; in the UK these demanded merely that the steering wheel moved less than 6 inches when the car hit a block of concrete at 30 mph. There was little incentive to make cars safer or to include better bodywork or airbags, for example. But by the mid-1990s over 45,000 people were being killed annually in road crashes across the EU. And attitudes were changing – led by the UK.

The Berkshire-based Transport Research Laboratory (TRL) was established by the UK government as a centre for road research in 1933 and eventually privatised in 1996. It had long campaigned for a tougher frontal crash test combining a higher speed of 40 mph with an offset barrier that concentrated the forces of the crash on less of the car's structure. Car manufacturers kept up a steady resistance to improving standards but with Department of Transport pressure, together with support from the Swedish government and the FIA, the TRL launched a project known as Euro NCAP. The first results were launched in February 1997 and aroused strong interest from the media. One of the cars tested, the Rover 100, collapsed completely and became an icon of the inadequacy of prevailing safety design.

Cars were awarded between one and five stars, with the Rover 100 unsurprisingly scoring just one star. Accompanying detailed reports specified the forces on the various parts of the test dummies used to represent car occupants and pedestrians. According to David Ward, a crashworthiness expert, 'the industry even tried to restrain their own competitive instincts and agreed not promote Euro NCAP results in their own marketing at all. This "gentleman's agreement" collapsed in a matter of weeks as more successful manufacturers couldn't resist boasting about a good rating!'

It's a competitive force that's resulted in stronger car structures, vastly better restraint and airbag systems and much improved pedestrian protection. Features such as electronic stability control (ESC) have been included in the assessment programme as well.

This is where computers detect skidding and loss of steering control. The system brakes wheels individually to help the driver regain stability, a technology compulsory on all new cars in the EU since 2014 and the US since 2012. Another area is the development of autonomous emergency braking (AEB), where cars detect objects in the vehicle's path and apply the brakes automatically if the driver hasn't spotted them.

David Ward claims that the combination of EU legislation and Euro NCAP crash ratings has saved around 78,000 lives since 1997. Today, 25,000 and not 45,000 people a year die in road crashes across the EU, which now has the safest road network in the world.

Safety gadgets of the future

Such improvements are only the beginning. To find out what novelties we're likely to see in coming years, I spoke to Matthew Avery of Thatcham Research, the centre established by the UK's motor-insurance industry in 1969 with the specific aim of reducing the cost of insurance.

He predicts that increasing numbers of sensors will soon be fitted to most new cars to detect vehicles you could collide with and even warn you about vehicles crossing your path. 'The next thing we're doing for Euro NCAP is Turn Across Path. A few cars, for example some Volvos, will already detect cars coming in the other direction and prevent you moving into the path of the other vehicle. The car uses the same sensors as it's using for emergency braking.'

Avery believes that a large proportion of deaths and serious injuries could be prevented by this technology,

Smart materials

Making cars safer has traditionally meant making them heavier, but this won't be true in the future. Materials will be more advanced, with greater use of aluminium, magnesium and composites arranged in ways to channel energies in a crash but without the extra lard.

More intriguingly, materials will also be smart and be able to change their state. One example of this is Google's sticky bonnet, or as they call it 'an adhesive vehicle front end for mitigation of secondary pedestrian impact'. Normally the bonnet has a conventional paint finish; but should the car strike a pedestrian the impact makes the surface adhesive. Basically, the person sticks to the bonnet instead of bouncing off and getting run over.

Apple has patented a shape-shifting car seat that can warn you of trouble ahead. New materials promise to be able to detect an impact and stiffen in a controlled fashion to absorb it. They'll be lighter and occupy less space, but be more effective at controlling a collision and preventing forces being transmitted to car occupants.

New materials may revolutionise roads too. Highways England is about to start trials of a new technology that allows sunflower oil to repair roads. The technique has been successfully tested by Dr Alvaro Garcia at the University of Nottingham, who was inspired by a Spanish edition of the television programme *Masterchef*. A contestant used spheres of jellified liquid in their cooking, which popped open and burst in the mouth, releasing their flavour – a technique known as spherification. In a brainwave of lateral thinking Garcia realised that capsules of oil could be placed in asphalt used for surfacing roads in order that when the roads start to crack the capsules would break open and release the oil within, softening the road around it. This helps the asphalt 'stick' back together, effectively filling in cracks and making roads self-heal.

including what crash investigators call Looked but Failed to See (LBFTS) incidents. This is where a driver, for example, wants to overtake and looks for traffic ahead but isn't actually concentrating. Their brain deceives them into thinking all is clear and they pull out on to the opposite side of the road. But all isn't clear.

'We can then start to do junctions as well. We're looking to introduce it into Euro NCAP 2020.' Cars will start looking sideways with special sensors fitted for the job. They'll be able to detect and alert drivers to avoid cars, motorcycles or children about to cross your path. They'll help prevent the sorts of horrifying crash seen in the 1970s Think Once, Think Twice, Think Bike campaign (see YouTube again). Rear sensors will also be included in the testing around this time. These will have an extended range to alert you to fast moving traffic steaming up behind when you're about to overtake. Autonomous Emergency Steering (AES) will go beyond current lane guidance, with its gentle 'nudges' on the wheel to keep you on track, and adopt a more assertive attitude, steering away from potential impacts.

The arsenal of airbags is set to expand even further in the near future. At present, side impact tests are focused on minimising the forces endured by the dummy sitting in the side of the car hit by the test barrier. Beefed-up body structures and airbags are designed to protect that occupant. But a third of casualties in side impacts are sitting on the opposite side of the car to the one that's been hit. For example, car occupants may collide with each other, banging their heads together. More airbags help to compartmentalise the car and assist in preventing grisly human-on-human carnage.

The Euro NCAP test will also start to pay attention to a tricky issue known as 'compatibility'. When cars are crashed into a static barrier – as is the case in most crash tests – they are really replicating a collision with another identical car. The reality is very different. Big SUVs hit small hatchbacks, sports cars slide under people carriers and, at the most extreme end, tiny city cars get squashed by trucks. As the car fleet is getting more diverse the compatibility problem gets worse. More extreme differences in weight and vehicle heights make it highly unlikely that crumple zones will crumple equally or that the 'right' parts of two cars' structures will coincide in a crash.

The result is an automotive arms race where responsible parents or those with a keen survival instinct buy the biggest vehicle they can. Volvo have been keen to point out that nobody's been killed in one of their jumbo-sized XC90s in the UK since the model was launched in 2002 – an achievement of invincibility that the model shares with the equally vast Audi Q7. But who knows how many superminis they've flattened.

Matthew Avery talked through how cars will change to compensate. 'It means you've got to make the Fiesta a bit stiffer and bit stronger and the Range Rover a bit weaker and softer so that the Range Rover is doing some of the work for the Fiesta.' The test will also encourage more uniform car structures with fewer aggressive hard areas that punch through other cars' bodywork in a crash.

Increasingly cars' restraint systems will be expected to adapt to different weights and heights of driver too. The sylph-like will have different airbag firing patterns to the big and burly. Older people currently have a poorer survival rate in crashes because they're literally more fragile than younger drivers. Seat belts for more mature drivers may need to be triggered more gently while smart materials that change their shape, strength or stiffness in response to a specific stimulus will also play a part in safety enhancement.

Euro NCAP is looking to prevent unattended children (and presumably pets) dying of heatstroke in cars by incorporating systems that monitor the temperature in the car and ventilate it automatically when necessary. Busy parents can apparently forget they've left their offspring locked in the car; and, as temperatures can reach 47°C inside when it's just 22°C outside, death occurs in minutes. Though not a huge killer, an average of thirty-six children die every year in the US this way and many more across the world.

Death rates are often much higher on the roads outside Western developed economies. This is consequent to multiple causes. In Thailand, where the death rate is ten times higher than it is in the UK, the large number of mopeds and motorbikes are a major factor. Elsewhere driving tests, road design and attitudes to risk are responsible. However, the statistics also reflect the fact that car manufacturers have tended to palm off older and less sophisticated designs on poorer countries – frequently charging top dollar in the process. Even where vehicles have been specifically designed for these markets they've been very poor in terms of crashworthiness. However, a Global NCAP program is now underway, devised and promoted by UK charity the Towards Zero Foundation, which campaigns for a world free from road fatalities. The program's mission is to bring European and US standards to the rest of the world, by naming and shaming manufacturers whose cars aren't up to scratch – the Rover 100 effect gone global.

In one round of tests in May 2016, all five India-made models achieved a zero-star rating – the Renault Kwid, Maruti Suzuki Celerio, Maruti Suzuki Eeco, Mahindra Scorpio and Hyundai Eon. But it looks like such shaming is having an effect. In 2018, the Tata Nexon became the first car made in and sold to the Indian market to achieve the Global NCAP's five-star crash-test rating.

Outside the orbit of NCAP, there are other safety developments expected. Car responsiveness and crashworthiness may have improved over the years, but it's become weirdly hard to actually see the road outside the car. If you throw yourself back in time and consider a boxy family car from the 1970s and 1980s, something like a Mk1 Golf or an Audi 80, you're immediately struck by the greenhouse-sized windows and your ability to see every corner of the car. It's amazingly easy to place on the road and park. But curvier, more aerodynamic bodywork with higher waistlines has reduced the glass area of more recent models. Similarly, the easiest way to meet demands for more crashworthy bodywork has been to thicken pillars, which has worsened visibility.

The Royal Society for the Prevention of Accidents in the UK thinks the problem is so bad that it has called for special driver training to deal with it. Cameras and screens might help, as well as transparent pillars with a lattice structure providing the strength. Volvo (who else?) showed such a prototype system as long ago as 2001 in its Safety Concept Car. Jaguar Land Rover and Bentley have used screens in the pillars as a more expensive and technologically sophisticated solution.

Meanwhile the HGV market, which is perceived as having a bigger and more urgent problem with visibility – sometimes resulting in collisions with vulnerable road users like pedestrians and cyclists – may lead the way in terms of legislation. In 2016, the London Mayor, Sadiq Khan, launched the world's first Direct Vision Standard (DVS) for HGVs. It rates what a driver can see through the cab's windows (not reliant upon screens, mirrors or cameras). If the plans are adopted, vehicles not meeting the required standard won't get an HGV Safety Permit and won't be allowed on London's roads. Similar standards are now being discussed across Europe as part of the European Commission's review of the General Safety Regulation. These are indeed positive advances, but it's no use having a better view of the road if you're looking at your phone.

Distraction driving

In several developed countries, road casualties have bottomed out and even started to rise again. As ever with car crashes, nobody's exactly sure why. However, the consensus is that it's down to our lack of concentration – and distractions. Whether it's the use of smartphones or dashboards practically built out of multiple touchscreens, there is more opportunity than ever to lose focus while driving.

Most of the evidence for distraction of professional drivers comes from naturalistic studies of HGV drivers in the USA, where researchers observe real behaviour while interfering with it as little as possible. One study reported that drivers were performing tasks unrelated to driving during 56.5 per cent of safety critical events. Also, drivers texting while driving were twenty-three times more likely to be involved in a safety-critical event than drivers who did not. However, drivers who talked on a mobile phone while driving were no more likely to be involved in a safety-critical event than those who did not. This would appear to give credence to the idea that it's looking down and interacting with a screen or keyboard that causes the most heinous levels of distraction, while holding a conversation with someone else seems relatively fine.

More careful analysis is required. There's long been quite a bit of uncertainty on the topic. Take road-safety charity Brake, which has come up with some rather odd suggestions over the years – including the idea of forcing newbie drivers to avoid motorways because of their higher speeds, despite the fact that they are far safer than other roads. Brake regards listening to music while you're driving as a no-no because it means you're using some of your brain capacity for listening. I think that's ridiculously puritanical and scientists seem to agree. The University of Groningen in 2013 suggested that, overall, listening to music while driving has very little effect on driving performance. Indeed, any effects that were measured turned out to be positive. Apparently music can help drivers to concentrate, particularly on long, monotonous roads. Personally I'm convinced I drive better while listening to the radio and I shall continue to inflict BBC Radio 3 on my passengers as an aid to my concentration.

There are a number of possible solutions to our distraction epidemic. These might include increased use of heads-up displays or wearables that reduce the need for manual interaction and therefore keep eyes on roads. Voice control sounds like another obvious fix but current evidence suggests it may be distracting in itself – maybe because voice control is so flaky. Another option is driver monitoring, whereby cameras keep an eye on our driving by recording our hand and eye positions. For example, some Volvos (them again...) display a coffee cup when driving becomes a bit erratic, which indicates that the driver should take a break. Cars could similarly use cameras to monitor whether the driver appears fully awake; if you look like you're getting drowsy or distracted an alarm would prompt you to get back in the loop. Some Cadillacs already warn you if they detect you're not paying attention.

From governments the usual response to all safety concerns is to lower speed limits and find more ways to enforce them. The European Commission has proposed that all new cars will need to have intelligent speed-assistance systems from 2022. These will inform drivers of speed limits and use the cars' cruise control to adapt to them. They would rely on optical observation of road signs and GPS and would enforce speed limits by slowing cars automatically.

GPS-based systems have been trialled for decades but have frequently proven useless because of inaccuracies inherent in the technology. For example, you might be stuck at 30 mph when joining a motorway because the system thinks you're on a parallel road with a 30 mph limit. Reducing a driver's options can be a dangerous game. There are also occasions where a quick burst of speed is necessary– like overtaking a long truck.

The autonomy question

The discussion so far has focussed on the incremental improvements in design and technology that have been happening for decades. However, when full automation finally arrives, all our safety fears may be answered in one fell swoop. Without human error casualties on the road will diminish to near zero – or so the theory goes.

In the meantime, we will have a range of semi-autonomous systems that might actually be more deadly than no driving assistance at all. Because drivers think they can relax and look

Lower speed limits – help or hindrance?

In France, road deaths fell to a low of 3,268 in 2013 but then started to rise again, reaching 3,469 in 2016. In 2016, 55 per cent of fatalities were on rural two-lane roads. So they decided to lower the limit on single-carriageway roads from 90 km/h to 80 km/h for a two-year trial period from 1 July 2018. The government was quick to credit the move with lowering deaths again (to 3,259). But road deaths actually started to fall at the start of 2018 before the speed-limit reduction was introduced. So, did the slow-down really address the problem or was it just a coincidence? Whichever is true, the policy emerged as one of the initial major grievances of France's *gilets jaunes* or yellow-vest movement.

In Denmark a similar speed-limit trial was reversed in 2017 because higher speed limits actually meant fewer accidents, maybe because people felt less need to overtake. Speed limits were also increased on certain motorway stretches and this also resulted in fewer fatalities on those stretches of road. The slowest drivers increased their speeds, but interestingly the fastest 15 per cent drove 1 km/h slower on average. The traffic flows better too.

In the United States speed limits on interstate highways have risen in recent years and researchers from the Insurance Institute for Highway Safety found that for every 5 mph increase in a highway's speed limit, roadway fatalities rose 8.5 per cent. The debate will continue.

away from the road, they can be lulled into a false sense of security. Tesla, for example, has been accused of using language that might encourage owners of their cars to think of the vehicles as self-driving when they're not.

In May 2016, Joshua Brown was driving his Tesla in so-called Autopilot mode in Florida, USA, and was killed when the car failed to distinguish between the bright sky and a large white eighteen-wheel truck and trailer crossing the highway. A report into the crash by the NHTSA was definitive: 'All evidence and data gathered concluded that the driver neglected to maintain complete control of the Tesla leading up to the crash.' Other crashes have followed, including those where telemetery recovered from the car shows that drivers weren't holding the wheel in the vital seconds or minutes leading up to impact. In some cases, drivers have admitted to having been engrossed in their phone or another task.

Arguably the most serious and pernicious form of distracted driving is with Level 3 autonomous vehicles. These are cars that can drive themselves but the driver must be prepared to intervene if the car signals it can no longer cope. Most of the time the car will run fine on its own – and that is where the ultimate problem lies. As Matthew Avery tells me, 'The danger is that people won't understand the limitations of the system. They'll say, "I can go to sleep and do so-and-so." If you know the car's going to steer for you, you'll get on with your email admin.' In other words, the more you experience the car taking control the more your confidence in it grows.

With fully self-driving cars the issue is often mixing human drivers with cautious autonomous technology on the same road. In February 2016, a Google Lexus SUV was driving itself down El Camino Real in Mountain View when it encountered some sandbags surrounding a storm drain blocking its path. It paused and then proceeded to move over a lane to pass them. A bus driver behind wasn't expecting the manoeuvre and collided with the automated car. The Google test driver thought the bus driver had spotted the car and would slow to allow it into the lane.

The first pedestrian to be killed by a self-driving car was Elaine Herzberg in March 2018. An experimental Uber vehicle,

An Uber self-driving car after a collision in 2017. Self-driving cars will have to be many times safer than human drivers to gain widespread acceptance.

operating in autonomous mode, struck and killed her while she was wheeling her bicycle across a four-lane road in Tempe, Arizona. Preliminary investigations suggested that the car's autonomous systems had detected Elaine six seconds before the crash but classified her as an unknown object. Only after several seconds delay was she categorised as a bicycle with an unpredictable path. Then, just 1.3 seconds before the crash, the car decided emergency braking was required. The car's automatic braking system, however, had been disabled to avoid erratic behaviour while under computer control.

Though there was a human driver on board to take over in emergencies, she was looking down at a display and failed to brake in time.

The incident caused a flurry of negative publicity and Uber paused their self-driving testing for nine months, before resuming in December 2018. However, self-driving tests have continued to increase in number and scale. Waymo has even begun testing without a safety driver behind the wheel in both California and Arizona. The vehicles may not be perfect but they are already pretty safe, with a lower at-fault accident rate than cars driven by humans alone.

The hacking threat

As more intelligent cars (hopefully) reduce collisions, they also introduce new dangers. Not least of all, there is the risk of hackers using vehicles as deadly weapons. This threat has already made it into mainstream movies like *Fast and Furious 8*, where thousands of self-driving cars are shown rampaging through New York, cascading out of control from multistorey car parks and hemming in the film's heroes.

At a technology show in 2006 I saw that Microsoft was promoting a new collaboration with Alfa Romeo to bring their Windows operating system to cars. The platform featured now-familiar developments such as Bluetooth phone connectivity, USB sockets for playing media and accessing files, and voice activation for weather forecasts. As well as these benefits it also seemed to bring potentially alarming aspects of the Windows experience into the car: viruses, the Blue Screen of Death and a reputation for hanging, freezing and software crashes, for example.

I can recall asking car companies and Microsoft at the time whether the dashboard version of Windows was fully isolated from safety-critical functions like braking, steering and throttle control. I was assured there was absolutely no interconnection. That may have been true then but it certainly isn't now. In 2015, two cybersecurity researchers, Charlie Miller and Chris Valasek, hacked into a Jeep Cherokee remotely using the car's Uconnect communication, navigation and entertainment system. The system proved to have no firewall or any other security that you might take for granted on your home computer so the researchers, once they'd got hold of the car's unique IP address, were able to take limited control of the car's self-parking system, operating its steering, engine and brakes.

An ethical hacker holds a circuit board from a Tesla. He'll try to reveal vulnerabilities and penetrate its security, finding what they call in the trade the 'attack surface'.

I went to visit Ken Munro of Pen Test Partners, an organisation employing several of the UK's leading ethical hackers – people who use their computer skills not to create criminal havoc but to expose weaknesses in systems for the common good. Ken is particularly passionate about car security and when I arrived the team was surrounded by a scattered assortment of Tesla parts. They were wielding oscilloscopes, laptops and other sundry gadgets. One person was hard at work attaching minute wires to a circuit board that would normally live in the Tesla's instrument panel, trying to find where electronic signals were unencrypted and what they could reveal about the car's security.

Historically the car industry has tended to regard security as a physical matter – best addressed by fitting tougher keys and locks. But with increasing numbers of in-car systems, protection against hacking is just as important. Munro takes a dim view of the industry's recent track record and cites what is known in the trade as the Megamos transponder hack. A group of researchers at Birmingham University and Radboud University at Nijmegen in the Netherlands discovered how to beat the transponder in the locking system fitted to a whole range of VW group cars, effectively leaving the vehicles with no security at all.

'When the researchers told VW about the problem and how to solve it, rather than VW engaging, talking to them and discussing with them, they slapped a non-disclosure agreement on them: a court injunction forbidding them from talking about it,' Munro says. VW chose to keep quiet and not issue a public recall. Instead they updated the cars quietly when they came in for service – a sort of soft-recall process that took about two years. During that time thousands and thousands of cars remained vulnerable.

Regarding the Jeep hack, Munro was scathing of how the company attempted to solve the issue: 'The "solution" was to post USB sticks to customers... By establishing the principle of sending USB sticks to an owner and the owner then updating it you're exposing them [to further threats]. All it would take was for a scammer to do exactly the same and no doubt one of them would have seen the exact syntax and format of the letter that came with the USB key. Owners could then be posted rogue updates and happily install them on to their vehicles.'

Hackers call the sum of the ways to get into a software environment the 'attack surface'. Such surfaces of most modern cars are both vulnerable and growing. Part of the problem is that the electronics of even new cars date from a bygone age. They're based round something known as the Controller Area Network or CAN bus. It was developed by Bosch in the 1980s and first appeared in a production car with the 1991 Mercedes S-Class W140. It's a messaging system that links together all the electronic-control units in a vehicle – the controls for everything from the engine management and the audio system to the power windows, lights and lane assist. It's robust and reliable but was developed in the pre-internet era and can be vulnerable to attack. Its central role means that the potential number of systems at risk in any hacking attempt is vast. The increasing number of cars that have the capability to phone home and connect to the internet only adds to the worries.

The E-call system was made compulsory in April 2018 on all cars sold in the European Union. It requires a cellular internet connection to let emergency services know a car's location in the event of a crash or other emergency. Penetrate the system and you can wreak havoc. In the case of an electric vehicle, even turning on the heating overnight could run down the batteries. Or, worse, you could inflict alarm and loss of control by suddenly turning on the sound system full blast when a driver is on the motorway. The opportunities for chaos are legion.

So far the Tesla model of security and updates seems to be the best. The security-conscious connected car can be updated remotely without trips to the dealer. Munro tells me that 'software updates can solve almost all security problems. Not all but most. That's where having a really robust over-the-air update platform is important. That's what most manufacturers are trying to do right now, so that when they do find a problem, instead of it being a multiple-thousand car recall which will cost millions and millions of pounds, they just push out a piece of software overnight.'

Car builders in the future will have to ensure they don't underestimate criminal resourcefulness and expertise, as they have in the past. In recent years, the whole keyless-entry saga has been one example of a woeful miscalculation. The problem caused a 20 to 30 per cent increase in car thefts in a matter of years. In the West Midlands area of the UK alone car thefts jumped from 5,344 in 2015 to 9,451 in 2017. Thieves can intercept the signal being generated by the key and boost it to gain access to the car and drive it away, even when the key is inside a house and a car is on the road outside.

According to Munro, 'People were aware of the problem about eight or ten years ago and the automotive industry was, as far as I'm aware, informed about it; but, because it was a very technically advanced hack, it was thought to be beyond the wherewithal of the average vehicle thief. The problem is that it doesn't take long for somebody to realise there's a market in this and automate the process.'

Attitudes in the automotive sector are slowly changing. Manufacturers are realising the huge costs involved if something goes wrong and they have to recall cars. But, as with all the Windows XP computers still in use today, it will be the older cars that are most at risk. What's even scarier is that when vulnerabilities are found they could apply to huge numbers of cars. Ken Munro asks me to imagine a vulnerability in all Range Rovers that could be exploited to make every one of them on the road turn left at the same time. It's a sobering thought indeed.

Even if hackers don't take control of your car they could try to extract some money by penetrating the infotainment system and installing ransomware. Imagine the first thing you see on a busy morning when you're in a rush to drive to the station is the satnav popping up with a message saying you can't start your car until you pay twenty quid. If the price tag was low enough and you needed to get somewhere urgently, you might well pay up. And the price would be higher next time.

Manufacturers may use network security as a further justification for forcing owners to use their dealers for servicing and repairs, rather like the ongoing iPhone saga where third-party screen repairs can stop the phone working after software updates. Only manufacturer-approved options will be allowed to be installed and we may see the end of common OBD (On Board Diagnostic) ports that allow independent garages to carry out repairs. Even dealers may not be secure enough to

Electric fireballs?

Tesla may have the best approach to online security but does electric power bring its own safety risks? Recalls for faulty batteries are a frequent occurrence in the gadget world. Samsung's Note 7 saga in 2016, with its unnerving catalogue of overheating and exploding phones, is one of the most notorious. The phones were recalled – twice in some cases – highlighting the potential fire danger of lithium-ion (li-ion) batteries.

Li-ion batteries in cars can suffer the same so-called thermal runaway, but their capacity is typically 8,000 times bigger than those in phones. They also suffer increased risks of impact damage. As the number of battery-powered vehicles increases, could there be a massive highway explosion waiting to happen? In a word, no.

In fact, electric cars could be less likely to catch fire than petrol-powered ones. Petrol is a risky material in itself and in the US there are about 174,000 car fires a year – equivalent to 55 fires per billion miles driven. Electric-car fires are relatively uncommon so far because there aren't many such vehicles on the road. Nonetheless, the latest estimates suggest that electric cars suffer fires at just a tenth of that rate.

The batteries are much better protected than the ones in a phone by firewalls and cooling jackets. Li-ion fires start slower than petrol ones, so there's more chance to get clear of a car in a crash. On the negative side, battery fires take longer to put out, can start days after a crash and can even reignite. When cutting bodywork to get someone out of an EV it's also vital to avoid high-voltage cables. Orange colour coding of anything over 60 volts helps prevent a very nasty shock.

be entrusted with the cryptographic keys that more complex connected cars will require. These may need to be kept at a company's HQ.

Fully autonomous vehicles will be much more vulnerable than cars today. The trouble with them from a safety and security aspect is their dependence upon arrays of sensors and vehicle signals that are so easily jammed. Signals without a physical connection can always be intercepted and wiped out by a criminal. Anything that requires vehicle-to-vehicle communications, such as platooning – where vehicles automatically follow one another in convoy – has safety vulnerabilities for which there is at present no known solution. In general, mass automation will bring with it a host of options for the tech-savvy terrorist.

And there's all that data to worry about

The threats from more intelligent cars don't have to result in huge crashes to be devastating. The vast volumes of data that vehicles are accumulating about their owners is another looming threat. For some high-profile users this could be a real security risk in itself. Even if your only issue is a colourful private life, a data leak might cause serious embarrassment. How secure are the car maker's data centres?

Current cars store quite a bit of data already, from phone contacts to satnav destinations. While this can be useful to potential future owners – a car's claimed low mileage might be questionable if the satnav shows regular trips to foreign holiday resorts, for instance – the present belief is that such data should be expunged from the car on sale. But routinely it isn't. Future owners connecting a car to an app or an online account can often see a whole litany of information. Former owners who stay logged in can also determine the current owner's whereabouts and journey history.

In Europe, the authorities are now wading into this emerging legal fog with GDPR, or General Data Protection Regulation legislation. This broadly prevents things being recorded. But it also conflicts with black-box legislation, which is set to mandate that all cars record their location, speed and state of various systems as they travel round.

As the stored data grows, questions arise as to who owns it – the manufacturer, the authorities or the user? The anonymity of data can easily be reversed by revealing locations and searches. It's going to be a pain if we have to agree permissions before every journey, even down to an individual-passenger level. I'm already sick of accepting cookies online. These requirements may be well-meaning but can be a right hassle if they're not properly thought through.

The incentive will grow for the individual user to hack their connected car to their advantage. At its simplest they'll want to jam or remove their black-box data or speed control. As autonomy becomes more commonplace there's the prospect that intelligent mobility systems will route self-driving vehicles in a way to collectively maximise traffic flow. But an individual user may find their car is taking a couple of minutes longer on a particular route. Rather like today's drivers chipping their engine control units (ECUs) to get a performance boost, clever users of the future might chip their cars not to follow the herd. When engineers started fitting big rubber bumpers to cars half a century ago, none of them would have imagined that data privacy and hacking might become the key issues in future car safety. It's a brave new world that might lead some to seek refuge in the more comforting one of old, familiar classics instead.

SIX

THE FUTURE OF THE PAST

Classic Cars and Enthusiast Drivers

Being nuts about old cars usually seems to mean getting together in a club with like-minded enthusiasts who share the goal of preserving them. The first outfit in the world dedicated to keeping classics alive was the Veteran Car Club (VCC), created in 1930 in the Ship Hotel, Brighton, and still in existence today. Its original aim was to conserve cars made up to the end of 1904, which were eligible for the annual London to Brighton Emancipation Run. This had begun in 1896 to celebrate an increase in the speed limit for so-called self-propelled vehicles to a heady 14 mph from a quite literally pedestrian 4 mph in the country and 2 mph in town. Later the club extended its scope to include 'Edwardian' cars made before 1919, as members became concerned that cars like Ford

Model Ts and Rolls-Royce Silver Ghosts were being scrapped at an alarming rate.

In 1934, a few years after the VCC was formed, the Vintage Sports Car Club was created, specialising in sporting cars manufactured before 1930. According to the mythology that's developed round its genesis, a group of enthusiasts were concerned that newfangled mass-produced cars were taking over from handcrafted traditional sports cars, which were considered to possess an artisanal beauty and mechanical directness. Or, in the words of Sam Clutton, who edited the VSCC's magazine, traditional cars created 'an intimacy between car and driver which is fostered by absolute positive and accurate steering, a close-ratio gearbox, aided only by a clutch stop and controlled by a sturdy vertical lever, and the (all-in-a-piece) roadholding that was characteristic of the best vintage years'. Although some of these vintage sports cars might have been only five years old in 1934, they needed urgent preservation before their joys were lost.

The club's still going today, with 7,500 members, and like the VCC now incorporates slightly more modern machinery as well. There is an alternative explanation for the club's formation that's based more on pecuniary matters than purity of design. Namely, the members feared that their cheap, old, second-hand sports cars could no longer compete with modern ones in races, so they needed to form a special club for their sluggish old bangers to do battle on equal terms. Nevertheless, it is the more romantic story that has stood the test of time.

Somebody loves them. The Reverend Colin Corke with his cherished Austin Allegro 1750 SS, finding pride and enjoyment in cars other people may not.

Every decade brings forth a crowd of passionate, misty-eyed enthusiasts wanting to forever preserve the cars' glories. I can recall there being a boom in interest in so-called classic cars during the 1980s, when drivers started appreciating just how magnificent some of the models produced in the fifties and sixties were – from Austin-Healeys and MGs through to Gullwing Mercedes and Jaguar Mk 2s. Now we have a 'modern classics' movement bent on preserving motors from the 1980s, the 1990s and later. And who can blame them. It's a bit of a golden age for me, including as it does the likes of Lotus Carltons, Golf GTIs, Lancia Delta Integrales, BMW M3s, Subaru Imprezas and Alfa SZs. I could name a hundred more I'd love to drive again tomorrow.

With the exception of the darkest days of winter, the British year has an astonishing variety of car events to fuel this passion. There are thousands if not tens of thousands of them, ranging from car tours and drive-it days to village fêtes and shows. There are social occasions that celebrate thoroughbreds, like the Goodwood Revival and the Festival of Speed, and the magnificent Silverstone Classic, where multimillion-pound cars are enthusiastically hammered flat-out round famous racing circuits while attendees dress in period garb. And it's not all top-end stuff. There's the rather politically incorrectly named Japfest, which celebrates recent classics from Japan, and there's my own favourite car show, the Festival of the Unexceptional, where cherished examples of the everyday cars of yesteryear are honoured, including those that were frankly a bit crap when still new. Recent winners have included a mint Nissan Sunny and an immaculate Austin Allegro 1750 SS, owned by a Birmingham vicar who once had twenty of the things.

Many of these events are anything but small-scale affairs. Over 350,000 people visit the Goodwood gatherings and classic cars have a huge following. In the UK alone there are over a million vehicles over thirty years old registered with the DVLA; and, according to a 2016 report by the Federation of British Historic Vehicle Clubs, an umbrella organisation for Britain's 400-plus old-vehicle associations, 8.2 million people in Britain are interested in historic vehicles. Historic cars are seen as part of Britain's heritage by 23 million people and the associated business employs 35,000 people and is estimated to be worth £5.5 billion to the economy.

There is similar enthusiasm elsewhere in the world. The Woodward Dream Cruise, held on the third Saturday in August in Detroit, is billed as the world's largest one-day classic-car event attracting 1.5 million people and up to 50,000 cars from around the globe. The emphasis is on muscle cars, street rods and American classics from the 1950s to the 1990s. Whole districts are turned into a dramatic historic carscape, like a giant *American Graffiti* scene made real. The Monterey Car Week is another top contender for every classic car enthusiast's bucket list. The Californian town hosts dozens of classic-car shows, culminating in the Pebble Beach Concours d'Elegance featuring 200 of the finest and most valuable cars in the world and an auction worth hundreds of millions of dollars.

Could the rapid change afoot in the motoring world threaten all of this? Well, arguably, it might if it makes owning and using a historic car more inconvenient.

Taking the MG for a spin

One of the most remarkable things about old cars is that they're so practical and easy to use. You can drive a 110-year-old Clément-Bayard on the same roads as a brand-new car and fill it up at the same fuel stations. The infrastructure is all still there. Slightly newer cars can even mix it perfectly easily with modern traffic. I can well remember the first time I drove a Rolls-Royce Silver Ghost and how straightforward and sprightly it was once I'd mastered the steering-wheel controls for hand throttle, ignition advance and mixture strength.

Most things in the heritage world aren't so refreshingly practical. Aeroplanes get grounded for safety reasons. Old electronics don't work, and are unfathomably difficult to repair. I can recall visiting the Victoria and Albert Museum a few years ago to assess the design of a computer with Kieran Long, who was in charge of the museum's digital collection at the time. Though many of the objects he was curating were less than twenty years old they had usually lost all their functionality, which he found hugely frustrating.

Most obviously disruptive of all is the fear that old cars will be regarded as too polluting for modern use. Old ordinary cars are set to get banned or heavily discouraged through taxes from city centres, so why not classics? We're used to having our roads open to all, but certain ages and classes of vehicle are increasingly likely to be excluded. Will this change our relationship with classic cars?

It probably depends on how tough councils are with the old stuff. Dutch cities initially seemed set on making no exemptions for historic vehicles in their city-centre, low-emissions zones. Earnest Dutch politicians regard all this very seriously – they don't want their good work on pollution reduction undermined by an army of classic-car enthusiasts. This has not necessarily gone down well. When a blanket ban on older petrol and diesel cars was proposed in Rotterdam there were mass classic-car protests that paralysed the city for a day in

Classic-car drivers on a protest ride in Rotterdam. They won't be a pushover for politicians touting disproportionately severe bans on historic cars in cities.

both 2016 and 2017. They worked. A historic vehicle exemption has now been allowed.

In Britain it looks like we may avoid such draconian measures and the associated protests. Britain's classic-car interest is centred around its plentiful old-vehicle clubs and the Federation of British Historic Vehicle Clubs is very proactive in looking after their interests. Contrary to what you might expect, this isn't an outfit dominated by enthusiasts with beards, tweed suits and a penchant for real ale. Their committee may be largely composed of gentlemen of a certain age but their number includes several ex-FTSE 100 board members who know their way around the corridors of power in Westminster and lobby very effectively if a threat looms on the horizon.

When I visited the Federation's communications director, Geoff Lancaster, at his Lincolnshire home I found him in his garage, actively maintaining his Formula Ford racing car and his classic Maserati. The Federation has successfully argued that historic vehicles should be exempt from the Ultra Low Emission Zones in London.

'Our argument was that the percentage of historic vehicles in a given area is extremely small,' says Lancaster. 'If every single historic vehicle which is registered within the postcode area of your emissions zone were to come out on the same day the additional load on the emissions in that area would still be immeasurably tiny. It's a disproportionate action not to exempt historic vehicles. They provide much more heritage value to the public than any damage they could conceivably cause.'

I noted his use of the term 'historic', which refers to cars above a certain age. The more commonly used term 'classic' is meaningless when it comes to serious matters like legislation. 'Classic means nothing at all in our world. It's the same as beauty. It's in the eye of the beholder. There's no definition. It's your opinion. Personal taste,' says Lancaster. There is some debate about how old 'historic' is, however. The EU's definition is thirty or more years old. The UK government, on the other hand, is adopting a requirement of forty years and older.

'London's an interesting precedent because it doesn't mean everywhere's going to do it. Basically the government has delegated how low emission zones are going to be regulated to a local level. But every time one comes up, we're going to be lobbying with local authorities. And we're going to use exactly the same argument.'

In 2017, the then Secretary of State for Transport Chris Grayling said that there were many big transport problems facing the UK, 'but classic vehicles are not one of them'. He pledged that while he was Transport Secretary he would not do anything to undermine owners 'getting out on the roads and enjoying using their classic cars'. Hopefully his successors will share this rational and wise view (probably, in many people's eyes, one of the few rational positions he has held). There are positive signs elsewhere in the world too. Paris was originally due for a ban, but now anything older than thirty years counts as historic there and will be allowed to be used, even after the 2024 diesel and 2030 petrol-car bans in the city.

Out of gas?

One practical issue that may become more irritating for classic-car owners in the future is the availability of fuel. So far fuel issues have been minor. The switch to unleaded was a problem for old cars in the 1990s but that's largely been dealt with by using additives or making engine modifications. There are also still a handful of stations in Britain that pump old-fashioned leaded fuel (at about £2 a litre). More recently the addition of ethanol to petrol has caused problems. It can attack the copper fuel tanks and soldered joints of certain veteran cars, but this problem is usually sufficiently minor to be solved by selecting super-unleaded fuel with its lower ethanol content.

Fuel could be a much bigger inconvenience if we're all running on batteries or hydrogen; there'll no longer be any handy roadside filling stations for your traditional motor. There will still be companies who'll deliver anything you like in a drum to your door, such as high-octane race fuel for your historic racer, or 'motor spirit' for your 1885 Benz with a surface carburettor that relies upon the volatile fuel's rapid rate of evaporation. But all this is possible only at considerable financial cost and barely practical if you're planning on using a historic vehicle regularly.

Such developments may end up sorting out the expensive classics from everyday ones. The owner of a pedigree Aston Martin or Bentley is likely to have the means to trailer a car to somewhere they can drive it in private and with supplies of fuel shipped in. But an everyday MG driver might find the requirement to have fuel delivered to keep in the garden or garage quite a disincentive to historic-vehicle ownership.

In a rather negative scenario the classic car of the future could be like the horse of today. Once a daily form of transport enjoyed by the masses, it may soon be a recreation enjoyed by only a tiny minority. It would no longer be used regularly on the road, but would instead be towed to the race circuit or leisure park by a modern or autonomous vehicle to be used by its relatively wealthy owner before being taken back to its cosseted home.

If petrol does become a rarely used specialist product, a more positive possibility is that concerns over pollution will evaporate along with any suggestion that we might eventually run out. In this respect the relationship between fossil fuels and cars reminds me of the one between silver and photography.

Back in the 1980s, as photography became more popular, there was a real fear that we'd run out of the precious metal, which plays a vital role in making traditional photographic films and light-sensitive papers. Instead digital photography became universal and film enthusiasts can now pursue their thriving niche without any fear of depleting global resources.

We might see more classics being converted to battery power. In the 1997 movie *Gattaca*, set in a future world where human genetic modification is the norm, people drive round in classics converted to use turbine-driven electric power. And a very splendid range of classics they are too, ranging from a Studebaker Avanti and Citroën DS Décapotable to a Rover P6 and a Buick Riviera. We're not at the stage of turbine conversions just yet, but battery conversions of classics are already quite common and could become more widespread in the future.

In the real world the development has always intrigued yet disturbed me – my normal enthusiasm for classics has crashed head-on with my passion for new technology. A very large part of the appeal of classic-ownership experience is that characterful mechanical engine, with its sound, feel and performance curve unique to each car, often mated to a manual gearbox for the satisfaction of gear changing and control. The whine of an electric motor and the gear-free transmission threatens to neuter and enfeeble the classic-car experience.

When my fellow *Gadget Show* presenter Jason Bradbury bought a DeLorean and announced his intention not only to convert it to full *Back to the Future* time-machine spec, with date displays, flux capacitor and all, but also to replace its (albeit rather feeble) Peugeot-Renault-Volvo V6 with an electric motor and batteries, I tried to dissuade him from the whole project.

Early conversions tended to use basic lead acid batteries giving short ranges of fifty or sixty miles and limited performance. But the motors have become faster and the battery technology more advanced, and a growing number of cars are being converted. The torque advantage of an electric motor has even been translated into drag-racing prowess.

A French firm called Ian Motion has begun converting classic Minis to lithium-ion power with a range of about 100 miles and a price ticket of around £35,000. In California, a company called Zelectric converts classic Porsches and VWs using lithium-ion batteries to give a similar range starting at $56,000. The VW Camper is in some ways the ideal conversion candidate. It's relatively plentiful so you're not doing anything sacrilegious to a slice of automotive heritage, and it didn't have a particularly brilliant engine in the first place. The performance improvement is vast and by careful placement of the batteries you even improve the somewhat wayward handling. Heating is taken care of with a supplementary electric-powered ceramic unit – though camping trips would have to be planned round your charging programme!

A key question in an electric-classic conversion is whether to keep the original gearbox. Enthusiast Steve Labib of Jozztek converted a Mazda MX-5 into a 350 horsepower car. 'With the torque these motors provide, there was no point driving it in

It's a drag

American electric-car gurus Mitch Medford and John Wayland have a passion for converting classics with electric powertrains. Their creations have posted frankly terrifying stats in drag races, far exceeding their petrol performances. The numbers speak for themselves, beating the acceleration of the fastest production cars available today. You can find them on YouTube.

	Standing quarter-mile	Terminal speed	0–60 mph
1968 Ford Mustang V8 289 3-speed manual (petrol)	16.3 sec	85 mph	8 sec
1968 Ford Mustang (Electric Conversion – Zombie 222)	10.24 sec	140.8 mph	1.94 sec
1972 Datsun 1200 Export Saloon (petrol)	19.1 sec	70 mph	14.5 sec
1972 Datsun 1200 (Electric Conversion – White Zombie)	10.4 sec	117 mph	1.8 sec

anything but fourth gear and it would spin its wheels in third. I really wish I'd taken the gearbox out, then put the motor in the gearbox hole and left the whole under-bonnet area empty.' He thinks old classics are easier to convert. 'On modern cars it's hard to make things like the antilock brakes or any of the electrical systems work without the engine in there. On an old car you find the power lead and connect it to an auxiliary battery and all the electrics work.'

I wouldn't be surprised if volume-production electric kits, which will bring the costs down, are developed for common classics like Morris Minors, MX-5s and MGBs. Sensitive cyborg-like conversions are already a small but significant part of the classic-car world and look set to become more popular in the future. This may be just as well, as the skills necessary to repair classics might begin to die out.

Where's my wrench?

Modern car-mechanic training schemes reflect the modern maintenance world, where you're as likely to be chasing fault codes on a laptop as you are to be dealing with greasy metal parts. With electrification and fuel-cell vehicles, not to mention increasing automation, the gap between some classics and modern cars is set to become even wider.

'A lot of skills are in danger of being lost – mainly through natural wastage.' says Geoff Lancaster.

'In the next ten years there aren't sufficient trained people coming through the system to replace those that'll be lost in the next ten years. We're not talking about people with sophisticated skills, like trimmers and English wheel operators [a device that helps shape complex sheet metal panels by hand]. Just basic mechanics. Take a kid going the Light Vehicle Apprenticeship, which is what the current franchises put their trainees through; you will not find the word "carburettor" anywhere.

'We're trying to replace the people being lost to retirement. We've started a historic-vehicle apprenticeship, which basically teaches all the traditional skills. And a lot of kids are leaving the Light Vehicle course to do ours. They do a year of Light Vehicle and some say they've never touched a car in all that time. They've looked at laptops and analysed diagnostic programs but what they actually want to do is pick up some spanners and get oily.'

The Federation's permanent training workshop has been up and running for the last four years at Bicester Heritage. The place is a fascinating microcosm of the future of classic cars – and maybe the future of enthusiast driving of the non-classic variety as well. Founded in 2013, it's based at a 350-acre former WW2 RAF Bomber Training Station, a location reeking of heritage and history before you even see an old car. It's a sort of classic-car hub that's already home to thirty specialist businesses and a track for driving a classic; soon there'll be a hotel and classic-car resort too. According to their website: 'The unique ecosystem of businesses acts as a "marina-like" cluster, promoting skills and expertise, employment, qualified customer footfall, shared business incubation and aggregate growth. Collectively, Bicester Heritage is all about driving the future of the past.' To me it looks like they're well on course to giving the past a thoroughly healthy future.

New skills will be always be needed. The classic-car passion is a moving target. Younger enthusiasts are interested in younger, more technological cars. A 1980s MG Maestro and its talking dashboard may not require the same skills as the immediate post-war classic MG TC. Talking dashboards were a quite a faddish technological dead end that appeared on a few models in the mid-1980s. The Maestro's was the most famous in Britain, fitted to the top-of-the-range MG and Vanden Plas versions of this Golf-size car that first appeared in 1983. Desperate to give the rather conservative design a technological edge, they ordered suppliers Smiths and Lucas to develop a digital dashboard complete with voice synthesiser. The result has a thirty-two-word vocabulary, recorded in fifteen languages, warning of potential troubles like 'high engine temperature' or 'battery not charging'. It also gave oral explanations for the display on the also newfangled trip computer, with phrases such as 'average fuel consumption' and 'trip distance'. A rotary volume-control knob is attached and the voice was provided by New Zealand-born actress Nicolette McKenzie, who had recently appeared in the BBC soap *Triangle*, set on a North Sea ferry.

Some of the earliest hybrids will soon be old enough for historic status and the best of them – like the Honda Insight, with its function-over-form clever design – are already in the frame to be future classics. Then there's the newer VW XL1, a tiny diesel hybrid that I remember felt almost like a Morgan 3-Wheeler when I briefly drove one round the roundabouts of Milton Keynes. Who knows, even the Toyota Prius, which has been hugely significant commercially yet never desirable from a driving perspective, might one day be considered a classic.

Future classics

Guessing correctly which cars are going to be future classics can be hugely profitable. But while it can yield a far better return on your investment than putting the money in the bank, it can be tricky to separate the winners from the losers.

Big middle-of-the-road saloons or humdrum hatchbacks aren't likely to be good bets. Relatively rare, high-power versions of models that look great and drive well are better choices. Convertibles often appreciate more than hardtops. Cars of the 1990s and early 2000s that are transitioning to become affordable modern classics include the Alfa GTV V6, Renault Clio V6 Sport, Jaguar XKR Supercharged and Nissan 350Z.

More expensive opportunities include the Porsche Boxster Spyder and recent stars like the BMW 1M Coupé. Some cars, like the Alpine A110, are so good you might want simply to buy them new and hold on to them. With others it's better to wait for that point where their value reaches rock-bottom and purchase a low-mileage, unmolested example just as all the rotten ones are heading for the scrapper. It's a tactic that's worked well in recent years with fast Fords and may do the same for a whole new generation of Focus RSs.

But are modern cars with all their ECUs and a whole host of other electronic equipment too sophisticated to keep alive as classics in the future? What about something as complex as a Mercedes S-Class? When it's pushing thirty, how are you going to keep ahead of failures in an equipment list that includes a perfumer, a 3D camera that links with the suspension, an ionised air-filtration system and seats that give a hot-stone massage? One car that regularly crops up on lists of potential future classics is the Audi TT. How easy are they to keep on the road as they enter the springtime of their glorious classic afterlife? I seem to remember that the TT, like the Triumph Stag, wasn't exactly a paragon of reliability in the early days. If anything they appear to be getting less demanding with age. According to Neil Crayford of Norfolk Performance Car Sales, a TT specialist, the only modern problem they have is a failure of the dashboard pod. This was also a common problem when the cars were new. But whereas an Audi dealer would have charged a small fortune to repair it back in the day, they can now be repaired for £120.

Perhaps I shouldn't be surprised. In my experience few people can be bothered to diagnose and repair a faulty control board on a kaput tumble dryer or oven, yet with cars the same components form a much smaller portion of the total value of the machine so there's much more incentive to find a solution. Whatever goes wrong, if there's a will to keep a classic car going, there'll be a way to do it.

Even airbags don't appear to be a problem. As the materials are tightly packed away from light they don't seem to deteriorate and cause problems. This is a hypothesis backed up by a US Insurance Institute for Highway Safety crash test in 1993 with a 1973 Chevrolet Impala, one of the first cars equipped with airbags. Though the car had done over 100,000 miles, and the radio and clock weren't functional, the airbags worked perfectly.

Can you still get the parts?

Parts availability is always a big worry for classic owners but if anything that's likely to improve in the future. For a start, if the car's manufacturer survives, they'll be taking more of an interest in bits for their old cars. Manufacturers are increasingly realising the profit potential in their heritage vehicles and the trend is likely to continue. You can, for example, get most of the parts for a 1950s Gullwing Mercedes from your local dealer, even including the seat upholstery.

In some cases manufacturers are actually still making complete heritage cars. Jaguar's Continuation series of lovingly re-crafted classic Jaguars is astonishing in its quality and attention to detail. The Jaguar Classic workshop has recreated six 'missing' E-Type Lightweights that were never produced back in the 1960s and nine £1 million Jaguar XKSS models that were originally lost in a Coventry factory fire in 1957, and is in the process of building twenty-five Le Mans-winning D-Types that complete the intended original production of 100 cars. Here, everything from the wheels and the gauges to the rivets and

The six £1 million E-Type Lightweight Continuation models built by Jaguar's Classic division are a magnificent time-shift legend within a legend.

the carburettors are as good as the original, and the cars take 10,000 hours of labour to make.

Usually such replicas aren't road legal because, as new cars, they would need to conform to the same current regulations as other new cars. But it's not cut and dried. Jaguar's six missing E-Types, for instance, have the potential to be fully road legal because they had originally been issued chassis numbers back in the 1960s but were deliberately sold in racing specification to encourage use on the track. Individuals building their own replicas in the UK can obtain 'Individual Vehicle Approval' for their creations. Though the four-hour test is tough it does not mandate crash testing, and the emissions requirements are those for the age of the engine fitted rather than for current ones.

With those cars that aren't deemed road legal, the fact that they have to be transported to tracks for use tends to increase the possibility of a future rich-versus-poor classic-car divide.

Porsche now makes parts for a huge variety of its old cars and is expanding its catalogue with 300 new components every year. Even arguably less prestigious models like the 924 Turbo are included. As Ed Myland from the Porsche 924 Owners Club told me: 'We're so lucky that Porsche is now supporting the car so much – there are 4,200 bits you can just go into a Porsche dealer and get.' The other way in which parts availability is improving is through 3D printing. This is bringing the price down and making an increased range of components readily obtainable. Again I found evidence at Bicester Heritage, where a company called Retrotech is dedicated to using the technique to help with the restoration of historic vehicles.

'Recent advances in laser scanning, together with digitally enabled manufacturing methods such as 3D printing and multi-axis machining, have greatly reduced lead times and tooling costs for one-off and short-run components,' says Retrotech chairman Francis Galashan. 'Parts that previously would have taken months of painstaking work by skilled pattern-makers can now be digitised within hours, then manufactured directly from a 3D model with big savings.' The company has helped make inlet manifolds for Metro 6R4 rally cars, exhaust manifolds for 1960s Sunbeam Tigers, brake parts for pre-war Alvises and a whole range of engine parts for vintage Bentleys, including the blocks themselves.

Even at an amateur level, the 3D printer is paying dividends. Ed Myland told me more: 'We have a couple of specialists in the club who are working on remaking parts, which is fantastic. The 924 Turbos have these special grilles on the bonnet which were fitted to nothing else; when they come on eBay they're invariably in very poor condition and going for £80 each. They're completely unavailable new so we're getting those 3D printed. Headlight washers are becoming difficult to obtain so they're being 3D printed as well. Even for pattern-making, if we're getting any metal parts cast, we'll use the tech where we can.'

It's possible that 3D printing could even ease the pressure on classics being crashed for entertainment. One of the more depressing aspects of seeing a historic car in a movie or TV series is that it's likely to meet some terrible fate. Think of Thelma and Louise's Ford Thunderbird, or the 1969 Mercedes 280 SE cabriolet in *The Hangover*. This is destruction you don't get with historic houses or paintings. In some ways the more expensive fare isn't the problem. There are numerous Lamborghini Countaches still in working order, so it arguably doesn't particularly matter as much when one gets destroyed in the making of *The Wolf of Wall Street*. Director Martin Scorsese wanted to use a real car rather than a model for more authentic bodywork damage in the crash sequences.

But often a lower proportion of the ordinary cars made survive so a film can have a more devastating effect on the remaining population. The Black Beauty car used by the Green Hornet in the 2011 film of the same name was based on a mid-sixties Chrysler Imperial, an ordinary car and a sort of American Austin Westminster. The production used twenty-nine vehicles, the vast majority of which did not survive the various stunts and action sequences. The producers of *The Fast and Furious* franchise are also often criticised for the number of classic vehicles they destroy in their action-packed movies.

The 2012 Bond film *Skyfall*, however, avoided having to harm a real Aston Martin DB5 by 3D printing several replicas at one-third scale. They were each assembled from eighteen components to give realistic door- and bonnet-opening functionality. One hopes to see such techniques more widely adopted in the future as charismatic classics get in shorter supply. If they ever remake the first *Godfather* movie, the pretty 1946 Alfa Romeo 6C driven by Michael Corleone in Sicily that gets blown up along with his bride won't be genuine, just a replica. With current values well over £100,000 there may be no choice in that.

The $100 million car

The most valuable classic car in the world – a Ferrari 250 GTO with racing history – changed hands privately for $70 million in June 2018. While a private purchase saves on auctioneers' fees, nobody would describe that as small change. A whole industry is making a very tidy income on the back of this particular asset class and it surely can't be long before we see a car hitting $100 million. Values have traditionally gone

The top ten most expensive classic cars sold at auction*

*Adjusted for inflation at 2018 prices

Date	Car	Price
26/08/2018	1962 Ferrari 250 GTO	$48.4 million
14/08/2014	1962 Ferrari 250 GTO	$38.1 million
05/02/2016	1957 Ferrari 335S	$37.3 million
12/07/2013	1954 Mercedes-Benz W196	$31.8 million
10/12/2015	1956 Ferrari 290 MM	$29.6 million
17/08/2013	1967 Ferrari GTS/4 NART Spider	$29.6 million
16/08/2014	1964 Ferrari 275 GTB/C Speciale	$27.9 million
19/08/2017	1956 Aston Martin DBR1	$23 million
25/08/2018	1935 Duesenberg SSJ	£22 million
19/08/2016	1955 Jaguar D-Type	$21.8 million

up and down as investors have entered and left the market in response to returns available elsewhere in the economy. Likewise, trends for individual models vary as tastes change. But over the whole market, prices have increased hugely in recent years. The Hagerty 'Blue Chip' Index averages the values of twenty-five of the most sought-after cars of the post-war era. It's seen prices increase nearly five-fold from January 2007 to January 2019.

One driver of such price appreciation is a newly powerful international force in the market: China. As Geoff Lancaster says, 'The Chinese love brands, and have an affinity for high-end historic vehicles.' He once asked the chairman of the Chinese equivalent organisation over. 'We took him to Prescott and let him drive a 1920s Bugatti Brescia up the hill. He was literally speechless for half an hour afterwards; it was the pinnacle of his experience. He'd probably never been near anything like that, let alone actually driving it. And at such an iconic location. We also took him to a Bentley Drivers Club meeting and he drove a 4-litre Blower Bentley that one of the members had brought. If he could have imported it, he would have got his cheque book out there and then.'

At present, the Chinese government doesn't allow the importation of historic vehicles. According to Lancaster, 'The rich Chinese guys all keep their cars in either New York, Paris or London. Some are in the city and others are in hangars with lots of Carcoons [the trade lingo for a plastic tent intended to protect your car while simultaneously ensuring adequate airflow]. They fly in on their Lear Jets; the car's already warmed up for them and they take it off on a scenic tour. They bring it back on Sunday afternoon, it goes back in the Carcoon and they fly back home.

'There's quite a bit of politics in the world classic-car movement. We were invited to go to China for a convention to convince their central committee. We made a presentation about the importance of heritage to the economy to try to convince them to liberalise the importation of historic vehicles.

Don't expect it to change overnight, because it takes decades rather than years, but they reckon, in the next ten years, it'll be liberalised and prices will go through the roof. There's such a pent-up demand and a huge purchasing power bottled up by legislation. Container ships will be thick with E-Types and Bentleys.'

The dark flip side to such burgeoning demand is that it means more incentive to manufacture fakes. Authenticity has long been a problem in the car world but with new markets it will get even worse, becoming more than ever like the art market. For example, take the MG Magnette K3, a short-chassis variant of the car that achieved great success in racing in the 1930s, driven by the likes of Captain George Eyston, Count Lurani, Tazio Nuvolari, Sir Tim Birkin, Whitney Straight and 'Hammy' Hamilton. Many ordinary K1 and K2 Magnettes were made into K3 replicas but their history is often confused.

Geoff Lancaster tells me that this can become tricky when the Federation of British Historic Vehicle Clubs has to rule on dating and provenance. One extreme position is where three people with three different cars are all asking the DVLA for an original historic number plate. All three of them can demonstrate that at least a proportion of the parts on their car are from the original but, over time, someone has turned one car into three cars. It's impossible to decide which car is the real deal.

There's long been a similar problem with Mk1 Ford Cortinas, whereby ordinary ones are tarted up and passed off as the Lotus high-performance variant – which featured a tuned twin-cam engine developed by Lotus, a vastly improved suspension, an interior bristling with extra dials, a repositioned gearshift and a wood-rimmed steering wheel. Increasing values means that the problem now affects a wider range of historic cars. Even the once humble Ford Escort is subject to the same treatment. There's growing demand for people who know their way round the intricate details that sort out the fake from the fortune.

As David Brown from Ford performance-car specialists RO Performance of Bury told me, 'If someone can find a two-door Escort shell they'll try and turn it into something better.' 'Something better' might mean the Mexico model, which had a 1,600 cc crossflow engine instead of the ordinary 1,100 cc and 1,300 cc motors, and was built to celebrate the model's greatest sporting victory in the 1970 London to Mexico World Cup Rally. You have to be a purist to know where the subtle changes are as they turn a five-grand car into one worth up to £50,000.

There are also genuine, openly manufactured replicas. Pur Sang is an Argentinian company that exhibits at the Goodwood Revival and manufactures replica Alfa 8Cs and Bugatti Type 35s. These are very, very accurate copies. They've got the patina as well. Pur Sang isn't passing them off as genuine but there's always the possibility that subsequent owners will. If the Chinese market restrictions come off, prices will go up, making the incentive to produce replicas even more irresistible. Given the Chinese fondness for making copies of ordinary modern cars, one wonders whether the classic-replica industry is about to get a kick-start.

Will classic cars be listed?

Discussions of art and provenance make one ponder this question. It's a difficult problem because, unlike buildings or works of art, cars are manufactured objects and not individual items. The Fédération Internationale des Véhicules Anciens (FIVA) is lobbying UNESCO for World Heritage status for classic cars. Geoff Lancaster says that 'through FIVA we've been trying for the last decade to get historic vehicles recognised as moving heritage in much the same way as they recognise ships and steam trains. There are hoops you've got to jump through. It's not a fast-moving organisation but the first thing you've got to do is declare a charter which defines what a heritage vehicle is, what condition it should be in and shouldn't be in. So we created a thing which was sent to UNESCO a couple of years ago called the Charter of Turin. Basically it's a manual of what constitutes a heritage vehicle.'

It's similar to the nonbinding but widely accepted Venice, Barcelona and Riga charters, which set standards for the definition, maintenance and preservation of historically significant buildings, watercraft and rail vehicles. The Turin Charter aims to 'support and encourage the preservation and responsible use of historic vehicles as an important part of our technical and cultural heritage'.

The definition UNESCO accepted is any vehicle capable of running on the highway under any form of motive power, including motorcycles but not bicycles. It's helpfully broad,

A work of art. The original Ford Mustang from the film *Bullitt* is on America's National Historic Vehicle Register.

including categories for cars, light commercial vehicles, trucks, ex-military vehicles and agricultural vehicles (perhaps surprisingly one of the biggest sectors of collected historic vehicles). Any that are thirty years old or more meet the definition.

In America, the Historic Vehicle Association is developing a National Historic Vehicle Register to carefully and accurately document America's most historically significant automobiles, motorcycles, trucks and commercial vehicles. In March 2013, it began collaborating with the US Department of the Interior to create a permanent archive of significant historic automobiles within the Library of Congress. There are so far only twenty-two vehicles that have achieved the accolade of a place on the register. They include the original 1947 Tucker 48 'Tin Goose' prototype, the 1968 Mustang that appeared in *Bullitt* and the 1938 Buick Y-Job, one of the first concept cars created by General Motors under the direction of their first design chief, Harley Earl.

Not all cars on the register are US built. There's also a 1954 Mercedes-Benz 300 SL W198 Gullwing Coupé. Apparently it's there because it was the first foreign mass-produced automobile built for and launched in America, and this particular one was owned by Briggs S. Cunningham, a prominent American businessman who achieved international success in automobile and yacht racing. It's also the first production car to use fuel injection and features a lightweight tubular space-frame chassis, an advanced aerodynamic body design and, of course, its distinctive 'gullwing' doors. Perhaps more surprisingly the register includes the 1985 Ferrari 250 GT California Spyder replica that appeared in the film *Ferris Bueller's Day Off*. This listing is based on 'its association with an important person and events in American history and culture'.

Are everyday classics the most endangered?

Most of the listings and talk of cultural heritage affect more expensive cars – the elite of the classic-car world. Lower down, the biggest danger might be a lack of interest. Is it the everyday classic that's most in danger?

One fear is that driving and car ownership as a whole will cease to be part of culture for younger drivers and potential classic-car owners. Cars may become inconvenient to keep in metropolitan areas and the classic car may become victim to this just like any other car. If you own a small hatchback, you can easily imagine driving a brand-new Ferrari. Will that still be the case in a world where the everyday experience of mobility is as a rented service? Will classic cars be seen as closer to motorbike riding or skydiving or hang gliding: something that's still pretty accessible but nevertheless a specialist activity?

Rob Symonds of the Morris Register encapsulates the challenge. 'The problem with classic cars in general is that you tend to warm to the classic cars you grew up with. I was a chemistry teacher at King Edward's School in Birmingham. I was there for forty-three years; I retired last year. I was into Morris 8s as a student. Younger generations are into the Escorts of the seventies, that sort of thing. Our membership increases year on year, which is not what a lot of clubs can say. But it's the demographics that are worrying. The average age of the membership must be solidly over sixty, and lots of our members are now in their seventies and eighties and they're going to the great car park in the sky.'

Some car clubs are attempting to reach out to younger owners by offering to lend one of their member's cars for a year, for free. The Morris Register decided to buy a car especially for the purpose and included free insurance as well. The club advertised the offer in various classic-car magazines. According to Symonds, 'It was like a lonely-hearts ad. Pre-war car seeks driver for a summer of love. Unfortunately we were a bit underwhelmed by the response. Maybe it's just too good to be true and that frightens people off.'

In the end the club selected a woman who lived in Derbyshire and taught art in Nottingham. She was very enthusiastic, apparently, and 'the fact she's a woman is a bit of a bonus as it is essentially a male hobby'. The trouble was that this 'young' driver was actually forty-four, which was older than they wanted. 'We hope there's some sort of equilibrium,' says Symonds, 'but the fact is it's getting harder and harder to get young people interested.'

Maybe he shouldn't fret. One conversation heartened me more than ever about the future of classic cars. The club that started it all, the Veteran Car Club, is still going strong. I spoke to its secretary, Stephen Curry, and he dispelled one worry immediately. 'I don't think there are many people alive who knew our sort of cars when they were new.' So much for people only loving the old cars of their youth, then. 'The old-car world has a life of its own.'

He did note a trend towards trailering cars to quiet areas of road where they can be enjoyed away from dense traffic. This process tends to favour smaller cars, which are easier to ferry about. So lightness is now prioritised over the need to keep up with modern traffic.

'Amongst American cars Model Ts were always looked down upon and people would go for Buicks and Cadillacs and things like that,' says Curry. 'Nowadays Model Ts are very sought after because they're light. You can easily put one on a trailer and tow it behind a normal car. Put a big old car on a trailer and try towing it behind a Land Rover and you find that it's even beyond the weight limit of that.'

The lesson from the VCC is that that the future of old cars is bright. 'There's still a demand, there is still an active membership. You could probably do an event with the Veteran Car Club somewhere in the country practically every weekend of the year. We're widespread and we're very strong.'

Personally, I think the future of the past, the destiny of our motoring heritage, is an area we have little need to worry about. A combination of leagues of enthusiasts, the rise of electric and hydrogen everyday cars easing the pressure on fuels and emissions, and the mainstream availability of new technologies like laser scanning and 3D printing all means our heritage will be savoured and appreciated well into the foreseeable future.

CONCLUSION

IS THE CAR DEAD?

We've seen how the car has been shocked out of its complacent stupor with better, cleaner power, added safety and ballooning intelligence-boosting convenience and comfort while cutting the casualty count. It's not the all-encompassing overnight revolution many had hoped for but it will nevertheless steadily transform our cars while retaining the joy of driving, keeping our ever-growing motoring heritage in rude health and satisfying our urge to race and compete. But could something more radical take the car's place? Some of the alternatives sound rather exhilarating.

In a sci-fi-and-superhero world there'd be no need for cars. We'd all use jetpacks. Yves Rossy showed the way. The veteran Swiss ex-jetfighter pilot turned 'Jetman' straps two carbon wings and four small jet engines to his body before swooping over the English Channel, looping-the-loop over the Grand Canyon or soaring over Burj Khalifa in Dubai. It's a trifle inconvenient that he has to throw himself out of a plane to get started but there's no denying his breathtakingly spectacular achievements.

He's not the only advocate of personal jet power. JetPack Aviation's $340,000 JB11, a pack with six kerosene-powered turbojet engines, more usefully takes off from the ground and it's claimed is capable of reaching 15,000 feet and 120 mph. Four of the engines are used to fly with two in reserve for emergencies. I found watching David Mayman fly up the hill with one at the Goodwood Festival of Speed in 2018 a genuinely thrilling experience.

Richard Browning, dubbed the 'Real-Life Iron Man', is another jetpack showman, from Salisbury in England. When I met him I was delighted to discover he's also a car lover with a 4.5 litre TVR Cerbera he uses for trials, drag racing and hill climbs. His rig has two micro-turbines belted to his back and two tied to each arm. My fellow *Gadget Show* presenter Ortis Deley was fortunate enough to try it. He found it frustratingly difficult. 'It's a weird feeling of lightness because you've never been lifted in that way before. It's not like being lifted under your arms. It's quite intuitive in the sense that you put your arms back to go forward. But I didn't get off the ground.'

That need for large amounts of operator skill and training, plus the vast cost and high noise levels, are sure to keep jetpack use restricted to the extremely wealthy in search of a few minutes of exhilaration. There are many other perpetually enduring fantasies that are never going to replace cars for very good reasons.

One of the most consistent – and consistently unattainable – dreams of personal transport is the flying car, the ultimate transport hybrid. Fiction and fantasy are full of them. A

London Transport poster of 1926, picturing the city as it would be in 2026, foresaw a sky full of car-like flying objects. Other examples are the Spinner police cars in *Blade Runner*, the Jetsons' Hovercar, Bruce Willis's flying taxi in *The Fifth Element*, Doc's DeLorean in *Back to the Future II* or even Chitty Chitty Bang Bang.

Even before these fantasy visions people were trying to build real flying cars. The 1917 Curtiss Autoplane is widely considered the first attempt. It featured large, non-folding but fully detachable wings (from a Curtiss Triplane) that proved rather too heavy and aerodynamically challenging for the 100 horsepower engine. The vehicle made a few short hops but never flew reliably. Other historic examples from the 1930s and 1940s include the Arrowbile, the ConvAirCar and the Airphibian.

More recently, five MIT graduates started Terrafugia in 2006 and commenced development of their Transition, with a price tag of around $280,000. At the time of writing none has been delivered to customers – not entirely surprising, given that it more closely resembles an awkwardly folding plane than it does a car. Nevertheless, the company, now part of the Chinese Geely Group, has already announced its successor, the TF-X, which is expected to debut in 2023. This looks like a far slicker piece of kit, at least based on the available 3D-rendered YouTube video.

So, the frightening expense of flying cars is prohibitive but they've also failed to get going because they're surprisingly inconvenient. Take-off normally requires a runway, which is rarely at hand. They also demand far too much skill from the driver – you need a pilot's licence on top of your normal driving licence. Above all they've tended to be a bit rubbish at being both a car and a plane. A car has to be strongly built for refinement and crash resistance, with large tyres to aid road holding. A plane needs to be light, with tiny wheels, and to get off the ground has traditionally required bulky wings, which rather get in the way when negotiating terrestrial traffic. The aerodynamics of the two are dissimilar and even the ideal drive systems for rotors and road wheels are vastly different. It's predictably almost impossible to design a single vehicle that's good at both.

The future of personal flying transport now looks more likely to be small electric vertical take-off and landing (eVTOL) craft and passenger drones. If anything, using the term 'flying car' puts you in danger of being perceived as a dinosaur. Uber, for example, has NASA on board to help develop its flying taxi, which you'll summon via an app and which will allegedly fly between the tops of tall buildings. Dallas and Dubai are set to get the first examples of this novelty.

Passenger drones won't be beating the morning rush hour for a while. Batteries aren't yet powerful enough for passenger-drone use. According to Uber, an air taxi will require batteries yielding at least 400 Watt-hours per kg but Tesla's most powerful batteries delivered just 250 in 2018. Another drawback is that drones take up far too much air space. Imagine everyone leaving an office block and hopping into their drone. The sky would be black with a dense swarm of buzzing activity. Also yet to be tested is whether the sensation of being whisked around in a drone is too close to a manic rollercoaster ride to be acceptable as transport.

The hyperloop is another idea that won't be threatening the survival of the car. The term was created by our old friend Elon Musk in 2013 while he was launching a competition to create a low-pressure tube that will enable passenger pods to zoom from San Francisco to Los Angeles in 35 minutes at 760 mph.

The concept itself is nothing new. British inventor and engineer George Medhurst took out a patent for trains operating in airless pipes in 1799. At the time they were more poetically described as 'atmospheric railways'. More recently there have been numerous plans for magnetically levitating trains in vacuum tubes. So far they've mostly been grounded at the pipedream stage by the cost of building a lengthy maglev system and the logistical difficulties of maintaining a near or absolute vacuum over long distances.

In Musk's twenty-first century version the air in the tube is at low pressure rather than in an actual vacuum, making the system less vulnerable to leaks. By virtually eliminating air resistance and the friction of wheels, high speeds can be achieved relatively efficiently. The tubes would be 6 feet in diameter and pods containing twenty-eight passengers would depart every thirty seconds.

Keeping air out of the tube is still a huge challenge. From Los Angeles to San Francisco the volume is 2 million cubic metres over 373 miles. The largest vacuum chamber in the world at present has just 1.5 per cent of this volume and requires enormous amounts of structural reinforcement. The steel tubes would need to withstand 10 tons of pressure per square metre and resist expansion and contraction with changes in temperature. Just one failure in a tube risks depressurising the whole system, potentially causing pods to brake suddenly from high speed as a wall of air enters the system.

If a pod gets stuck the ensuing pod will need the ability to make an emergency stop, so the distance between pods needs to be vast, limiting capacity. A slight failure in a pod's structure would cause the occupants to be exposed to a fatal lack of pressure – a death rather like being thrown out into deep space without a suit.

To cap it all, the nausea may be insurmountable. The hyperloop pods won't be doing a constant speed in a straight line. They'll be going uphill and downhill, round obstacles and doing stop-starts at stations. G loadings seven times higher than the limits within which Japan's Shinkansen bullet train operates have been suggested. The proposed $7.5 billion cost of the LA to San Francisco hyperloop seems short of a nought or two, or maybe even three. As vanity projects, perhaps, some hyperloops may be built but they'll rival trains and planes, not cars.

At the opposite extreme, there has been a naïve hope that, in urban areas at least, we'll all be whizzing around on miniature electric scooters. From personal experience I have to say these things have such miniscule wheels that they get caught in the tiniest potholes, launching you headfirst into oncoming traffic. Nevertheless, Silicon Valley has been enthusiastic and the most famous start-ups, Bird and Lime, have been valued in the billions of dollars. The idea is that apps track scooters' location and who's using them so you just pick one up and go, and teams of roaming personnel will find them when the batteries are flat and whisk them off for recharging. Supported initially by brainlessly optimistic politicians, they've become renowned for creating urban litter with abandoned devices – and for a series of nasty crashes.

A day after they were made legal in the Swedish city of Helsingborg, a rider was killed and the Swedish transport authorities immediately called for the devices to be banned. In Portland, Oregon, e-scooters are involved in accidents forty-four times as often as motorbikes, while in Austin, Texas, one in three users are injured on their first trip.

When it comes down to it, there is nothing that can compete with the convenience and cost-effectiveness of the car. The idea of travelling round in a box with (usually) four wheels on the ground isn't going away. Though city centres will become even more car-hostile and congestion may continue to get worse as the car remains a victim of its own success, there is nothing to indicate its imminent demise.

The dream of an autonomous, connected, shared and zero-emission future will live on but don't believe all the hype. We're not on the verge of a sudden and dramatic revolution. Instead, we'll see a more diverse and gradual evolution of the car.

The furthest away from manifestation is the nirvana of autonomy. The dream of a luxury transport capsule that appears at your house and scoots you off quickly to your destination, disappearing from view at the end of your journey, is a long way off. Artificial intelligence simply doesn't seem capable of providing it yet, at least without massive investment in infrastructure. A world of exciting targets has become one of missed deadlines. While automated pods look destined to be a slow form of transport in very controlled spaces, it will be some time before they make tentative steps further afield with special lanes on motorways and dedicated zones in towns and cities.

Zero-emissions, or something close to it, is a dream that we will attain earlier. Battery-electric cars will be most successful at either extreme of the car market. Some drivers will keep an additional electric city car for popping into or around town, while the rip-roaring acceleration of all-electric hyper-cars will appeal to rich ex-petrolheads who'll be queuing up and endlessly debating the merits of superstars like the Tesla Roadster 2 (1.9 seconds 0–60 mph) and Rimac C Two (1.85 seconds 0–60 mph).

Don't expect the roads to be entirely dominated by the whine of battery-powered electric motors, or whatever alter-native soundtracks have been installed to help you hear their approach. Internal-combustion-engine cars in the form of cleaned-up diesel and hybridised petrol models will continue being the mainstay of personal transport for at least the next couple of decades, and becoming more and more efficient at powering lighter cars made of more advanced materials. The lightweighting expertise of the likes of McLaren, with its clever composite confections, will spread down to a wider market. In the longer term hydrogen will join batteries to make an anxiety- and carbon-free choice for all types of driver. Meanwhile, our cars will continue to become more connected, contributing enormously to safer driving and, hopefully, increasing speeds again.

Car use and buying patterns won't change overnight either. Americans will still buy pickup trucks. We'll continue to dream of faster and faster cars and make more and more gorgeous designs. The future will have its own Lamborghini Miuras, Jaguar E-Types, Aston Martin DB4 GT Zagatos and, for that matter, Minis and 2CVs. Expressions of individuality and customisation will soar, aided by 3D printing. Our love of speed won't be put back into the box. Young people won't give up on driving, however much older people might wish them to. Indeed the very promotion or subsidy of car sharing, and attempts to cut car ownership by nanny-ish planning and punitive taxation, might actually make owning a car seem even more desirable and prestigious.

When Nikita Khrushchev, leader of the Soviet Union, visited the United States in 1959 he was appalled by the numbers of cars. 'In our country cars will be used more rationally than in America,' he declared. 'Taxi pools will be widely developed where people will obtain cars when they need them. Why would one rack one's brains over where to put the car, why be bothered with it?' This taxi-sharing model of personal transport never took off. Every Russian wanted a car of their own and spent years and years saving up for a rotten Lada or similar piece of junk.

There is a revolution ahead but it's not the autonomous and electric tunnel vision one might expect. The future is one of more intelligent, more diverse driving experiences. We'll embrace a much wider range of cars in the future and use more rather than fewer of them. We might cherish something historic and have the option to exercise and maintain it on a heritage campus, use a battery-EV for urban trips, and take something more autonomous with a hybrid or hydrogen power unit for longer journeys. We'll pay subscriptions to motorsport driving-experience parks where AI and VR offers unparalleled instruction on how to drive better, and witness new, more exciting and more involving forms of motorsport. We'll continue to experience the sensual delight of tyres straining against tarmac and cars like Lotus Elises, Porsche 911s and Ariel Atoms with chassis expressly designed to capture the sheer pleasure and dynamic thrill of tackling the twisting blacktop.

Perhaps a more gradual process is a blessing in disguise. We like to be in control. Cars give us ecstatic thrills while we sit down in a comfortable environment and listen to the entertainment of our choice. They represent freedom combined with some of the best tactile pleasures and sensations that life can offer. They're not going away anytime soon.

IMAGE CREDITS

Main Text

© Wikimedia Images, pages 16, 126 (top left, middle left, bottom left)

© Nvidia, page 26

© Waymo, page 35

© Tesla Motors Inc, page 50

© Laura McFarlane, page 69

© Riversimple, page 75

© Getty Images, pages 94, 102, 165, 167

© Jon Bentley, pages 108, 120, 122, 136, 201, 213

© Aston Martin, page 132

© Porsche, page 126 (top right)

© Land Rover, page 126 (middle right)

© Daimler, page 126 (bottom right)

© Ariel Motor Company, page 151

© Hankook Tires, page 175

© NCAP, page 184

© AP Images, page 198

© www.andysaunders.net, page 180

© Gert van der Ende, page 217

© Jaguar, page 230

© Ford, page 239

© McLaren, chapter opening illustration

Plate Section

Page 1 (both images): © Christofer Sætrang

Page 2 (top): © Shyamal Kansara

Page 2 (bottom): © Stavros Mavrakis

Page 3 (both images): © Getty Images

Page 4 (both images): © Mercedes

Page 5 (both images): © WayRay

Page 6 (both images): © Terrafugia

Page 7 (both images): © Renault

Page 8 (both images): © BMW

INDEX

Page numbers in *italic* indicate illustrations

Index